Procedures and Standards
for a Multipurpose Cadastre

Panel on a Multipurpose Cadastre
Committee on Geodesy
Commission on Physical Sciences,
Mathematics, and Resources
National Research Council

NATIONAL ACADEMY PRESS
Washington, D.C.   1983

Library of Congress Cataloging in Publication Data

National Research Council (U.S.). Panel on a
  Multipurpose Cadastre.
  Procedures and standards for a multipurpose cadastre.

  Bibliography: p.
  1. Cadastres—United States.   2. Real property—
United States—Maps. I. Title.
HD205.N37   1982   352.94'19   82-24557
ISBN 0-309-03343-8

Available from

NATIONAL ACADEMY PRESS
2101 Constitution Avenue, N.W.
Washington, D.C. 20418

Printed in the United States of America

Panel on a Multipurpose Cadastre

MacDonald Barr, Lincoln Institute of Land Policy, Cambridge, Massachusetts, *Chairman*
John D. McLaughlin, University of New Brunswick, Canada, *Cochairman*
Richard R. Almy, International Association of Assessing Officers, Chicago
Kurt W. Bauer, Southeastern Wisconsin Regional Planning Commission, Waukesha
Kenneth J. Dueker, Portland State University
Earl F. Epstein, University of Maine, Orono
G. Warren Marks, The Pennsylvania State University
Kenneth Strange, Turner, Collie & Braden, Inc., Houston, Texas

Liaison Members

John Behrens, U.S. Bureau of the Census
Charles Finley, National Aeronautics and Space Administration
Clif Fry, U.S. Geological Survey
Armando Mancini, Defense Mapping Agency
Jerome Smith, U.S. Department of Housing and Urban Development
James Stem, National Oceanic and Atmospheric Administration
Douglas J. Wilcox, Bureau of Land Management
Gene Wunderlich, U.S. Department of Agriculture

Staff

Hyman Orlin, *Executive Secretary*
Penelope Gibbs, *Project Secretary*

Committee on Geodesy

Byron D. Tapley, The University of Texas at Austin, *Chairman*
MacDonald Barr, Lincoln Institute of Land Policy, Cambridge, Massachusetts
Charles C. Counselman III, Massachusetts Institute of Technology
Adam Dziewonski, Harvard University
Edward M. Gaposchkin, Lexington, Massachusetts
John C. Harrison, University of Colorado
Buford K. Meade, National Oceanic and Atmospheric Administration (retired)
Richard H. Rapp, The Ohio State University
Fred N. Spiess, Scripps Institution of Oceanography

Liaison Members

John D. Bossler, National Oceanic and Atmospheric Administration
Frederick J. Doyle, U.S. Geological Survey
John R. Filson, U.S. Geological Survey
Bernard Hostrop, Bureau of Land Management
Armando Mancini, Defense Mapping Agency
Jesse W. Moore, National Aeronautics and Space Administration

Staff

Hyman Orlin, *Executive Secretary*
Penelope Gibbs, *Project Secretary*

iv

Commission on Physical Sciences, Mathematics, and Resources

v

Preface

In *Need for a Multipurpose Cadastre* (Committee on Geodesy, 1980) it is stated, "there is a critical need for a better land-information system in the United States to improve land-conveyance procedures, furnish a basis for equitable taxation, and provide much-needed information for resource management and environmental planning."

That report discusses existing land-information systems and the multipurpose cadastre as a basis for a dynamic, public process that efficiently collects, maintains, and disseminates land information. It not only identifies the land-resource-related problems faced by public and private organizations but also outlines the nature of a multipurpose cadastre as a means to remedy these problems. However, the questions of how governments, especially local governments, can carry out the recommendations made in that report were not answered.

To address the questions left unanswered by its 1980 report, the Committee on Geodesy of the National Research Council undertook this study on recommended procedures and standards for a multipurpose cadastre. The report was prepared by individuals who have practical knowledge of land-information needs and problems at the local level and who have been active in efforts to satisfy those needs and to solve those problems, including members of university faculties concerned with these matters.

Contents

Procedures and Standards
for a Multipurpose Cadastre

Executive Summary

This report reaffirms the statements in the Committee on Geodesy (1980) report regarding the need for a multipurpose cadastre at all levels of government in the United States and suggests the outlines of procedures and standards that will be required for its design and implementation. It is intended to assist both the local governments wishing to pursue the development of cadastral records systems for their own counties or equivalent districts and also the many other regional, state, and federal agencies, as well as private businesses, whose participation will be needed.

The basic components of a cadastre are the following:

1. A spatial reference framework consisting of geodetic control points;
2. A series of current, accurate, large-scale base maps;
3. A cadastral overlay that delineates all cadastral parcels and displays a unique identifying number for each of them; and
4. A series of compatible registers of interests in land parcels keyed to the parcel identifier numbers.

In a *multipurpose* cadastre, these components must be maintained in a manner that provides the foundation for other registers of land data, each keyed to the standard parcel identifiers for retrieval of specific records and for linking with data in other files. The other files may be in the same jurisdiction or in any other governmental unit that has a multipurpose cadastre system.

Requirements for the *geodetic reference framework* for a cadastre are summarized

as: "First, it must permit correlation of real property boundary-line
…ographic, earth science, and other land and land-related data. Second,
…ermanently monumented on the ground so that lines on the maps may
…ced in the field. . . ."

…commend that the State Plane Coordinate Systems be used as the basis of
the multipurpose cadastres in each state. Monumented points of known location on
this system should be distributed throughout the area served, at intervals no greater
than 0.2 to 0.5 mile in urban areas and 1 to 2 miles in rural areas.

Only a handful of the more than 3000 counties of the United States currently
maintain a geodetic reference network with a density adequate to support a multipurpose cadastre. Indeed, in only about 10 percent of the 500 counties designated
by the U.S. Department of Commerce as "leading" counties in terms of economic
activity is there in place an existing primary geodetic framework of sufficient density
(spacing of 3 to 5 miles or less) to serve as the starting point for further densification
to a level that would support a cadastre.

Significant progress toward establishing multipurpose cadastres thus will require
extensive programs of densifying the geodetic control network. Fortunately, several
new technologies (described in Chapter 2) for accurately determining the positions
of survey control points promise that substantially lower costs per control point will
be realized in projects that are organized on a large enough scale to employ them.

The *base map* of a multipurpose cadastre is the primary medium by which
cadastral parcels are related to the geodetic reference framework; to major natural
and man-made features such as bodies of water, roads, buildings, and fences; to
political boundaries; and to each other. The base map also provides the means by
which all land-related information may be spatially referenced to cadastral parcels.
It is the medium for determining and expressing locations in continuous space, so
that shifts in the locations of the boundaries of cadastral parcels may be entered as
necessary in the official records. The map may be stored either in graphic form, on
paper or Mylar, for example, or in digital form as a "virtual" map.

Base maps should be prepared to meet United States National Map Accuracy
Standards (see Appendix B). Customary map scales for each type of area (urban,
suburban, rural, and resources regions), which are in almost universal use today,
are listed in Section 3.4.

The *cadastral overlay* depicts positions of property boundaries in relation to the
other features shown on the base map and shows the standard identifier of each
parcel, the latter serving as the key to the many other parcel records that can then
be based on the multipurpose cadastre. The cadastral overlay could be viewed as a
property ownership map that adheres to standards for accuracy of plotting of property
boundaries and completeness in display of parcel identifiers—including standards
for timely updating to show boundaries and identifiers of newly created parcels.
Although the boundary plotted on these maps should meet map accuracy standards,

it normally does not provide the legal description of the boundary—for the latter, the record of the cadastral survey normally must be consulted.

Accuracy standards for the land surveys that support the cadastre should be expressed in terms of boundary tolerances (maximum probable error, in feet or meters) rather than the traditional boundary survey misclosure ratio (e.g., such as 1 part in 10,000). User requirements for cadastral survey accuracy have not yet been clearly established. One recent study for the Maritime Provinces of Canada recommended maximum boundary tolerances of ± 0.1 ft in urban areas, ± 0.3 ft in suburban areas, and ± 1 to ± 2 ft in rural areas (see Section 4.2.2).

A wide range of governmental functions can benefit from use of the multipurpose cadastre as a complete inventory of all currently existing parcels and their legal identifiers, permitting convenient exchange of data among the users. Twenty-five such functions are listed in Section 5.1.1. The data requirements of the three functions that are predominant users of such data systems—property tax assessment, deed recordation, and planning—are described in some detail.

Exchange of land data between systems describing natural phenomena, on the one hand, and cultural phenomena, such as attributes of land parcels, on the other, is greatly facilitated when both are built upon the foundation of a multipurpose cadastre. Within the continuous space that is defined by a cadastre, the boundaries of natural areas can be plotted and compared with those of culturally defined areas. However, most existing natural area data have been compiled without the benefit of this accurate spatial referencing. Until their boundaries are referenced to the same coordinate system in the future, such data exchange will require arbitrary apportionments between natural and cultural areas.

The focus of the activity of organizing and operating a multipurpose cadastre will be in the offices of the county government or, in some areas, the municipalities that carry out the equivalents of what are county functions in most states.

We recommend that a central office in the government of each county (or municipality, where appropriate) be assigned the responsibility of managing the development of the systems of maps and files that will constitute the multipurpose cadastre for that locality and of compiling the common set of standards for definitions of data elements, accuracy, frequency of updating, and completeness of the records. To assure compatibility these standards should be developed in cooperation with other jurisdictions, including state and federal governments.

However, few county governments by themselves have had sufficient resources or the long-range political commitment required to develop a multipurpose cadastre. Assistance from several other sources will be needed, based on their prospective use of the output of the system. Support from state governments will be essential, specifically in (1) organizing the land-records function in county government, (2) mandating the support of a compatible system by units of state and local government and by the private utilities, and (3) providing financial assistance.

We recommend that a program of federal grants to counties (or their equivalents) be established to provide between 30 and 50 percent of the cost of developing multipurpose cadastres that meet or exceed federal requirements, subject to participation of the state government in the design and partial funding of the program.

The cost of a nationwide program of federal financial assistance is estimated at $90 million per year over a 20-year period. High priority should be given to drafting a plan for federal assistance, with a better projection of costs.

Support from federal agencies will be important in many aspects of a national program to develop multipurpose cadastres, including the following:

- Extension of the network of first- and second-order geodetic control points, to provide this basic framework in every county of the United States.
- Completion of the geodetic framework for the cadastre along the boundaries of federal lands, which in the 30 states covered by the Public Land Survey System will mean retracing, remonumenting, and determining the positions of all quarter-section corners along the boundaries of the federal lands, and in the interior of federal lands, where appropriate, with reference to the State Plane Coordinate System.
- Research and drafting of proposed standards for those components of a multipurpose cadastre for which federal agencies have established expertise, working in conjunction with the national associations of state and local governments in these fields.
- Requiring compliance by federal agencies and their grantees and contractors with the standards established nationally for large-scale cadastral mapping and cadastral data-base systems or, until such standards are adopted, with the relevant state-level standards.

Multipurpose cadastres will be realized in the United States as much by the coordination of investments currently being made in large-scale mapping and land-parcel records, pulling together federal and state as well as local interests, as by increased funding of these activities. Therefore, in conclusion:

We urge the National Association of Counties, through its appropriate constituent organizations and staff, to organize a review of the findings and recommendations of this report, involving representatives of local user agencies, and to identify the areas in which more specific standards and procedures are most needed to make the approach described here operational.

Examples of cadastral records programs are described in Appendix A to illustrate approaches that seem to be succeeding. They are not the only examples, nor necessarily the best, but nevertheless represent the procedures recommended in this report. These programs are characterized by their commitments to record the locations of property boundary corners with reference to a geodetic framework provided by the State Plane Coordinate System and to maintain high standards of quality in building their systems of land records.

These case studies are in distinct geographic regions of the United States: East, Midwest, and West. All are in relatively early stages of planning and testing of a records system, although the program in Wisconsin has made substantial progress toward completion of the geodetic reference network. Two of them represent initiatives by regional planning districts, which then depend on county-level governments actually to build and maintain the records systems.

Each of the programs also has a number of individual attributes that are exemplary. The program of the regional planning district that includes Milwaukee (Appendix A.1) has been closely tied from the beginning to the processes of surveying and recording of new property boundaries. The program in a suburban county west of Chicago (Appendix A.2) is being managed by the county executive office to support county-operating agencies. The program in the suburban county that adjoins Denver (Appendix A.3) has used the subdivision control process to enlist the resources of land developers in building the system of monumented property lines and the records that locate them with respect to the State Plane Coordinate System. The program of the regional planning district that includes Philadelphia (Appendix A.4) has included the electric and gas companies in the consortium that will develop and use the integrated system.

1
Introduction

1.1 PURPOSES OF THIS REPORT

This report is intended to provide guidance to local officials who seek to establish elements of a multipurpose cadastre in their jurisdictions and to the state and federal agencies that seek to assist in this effort.

The specific objectives are as follows:

To provide descriptions of public systems for cadastral products and services that can be permanent and cost-effective in the United States and references to sources of detailed specifications for development of components of these systems.

To encourage commitments by state and local governments to the development of multipurpose cadastres to serve their respective areas but without attempting to prescribe their administrative organization, which will need to be adapted to the existing governmental structure in each locality.

To suggest guidelines for federal and federally supported programs that can have an important impact on the development of multipurpose cadastres.

To encourage the adoption of compatible standards and procedures for the components of a multipurpose cadastre and related records.

Prospective users of this information include the following:

Administrators of local programs needing technical guidelines for program development.

Legislators, especially those who are looked to for expertise and leadership in matters affecting municipal administration and real estate.

6

Administrators of federal programs that provide technical or financial support for operations that are important in the development of a cadastre.

Professional planners, environmentalists, surveyors, lawyers, members of university faculties, engineers, and others in a position to provide leadership for establishment of cadastres.

The land-information officers proposed earlier (Committee on Geodesy, 1980) for designation by state and local governments.

1.2 SCOPE OF THIS REPORT

This report follows the lead of the Committee on Geodesy (1980) report, which recommends that local governments be the primary access points for local land information and that they maintain land data compatible with a multipurpose cadastre and transmit these data to higher levels of government when needed. Federal agencies can play an important leadership role by making their land-information systems consistent and compatible with each other to facilitate joint use of the data. Standards to assure this compatibility might be used as models by state and local governments.

With the wider availability of geodetic and mapping data in metric units, multipurpose cadastre development programs should consider the use of these units for all their elements and products. We recognize the tremendous costs that could be incurred in a program of conversion from existing English units into metric units. Hence, as initiatives that lead to the development of the various elements of the multipurpose cadastre are of major significance for this nation, these initiatives should not be deterred in any locality by the need to convert to the metric system. However, we would encourage that all cadastral data-management systems be designed to handle either English or metric units and that, whenever possible, metric products be made available.

1.2.1 Suggested Local Procedures for Building a Modern Cadastre

This report identifies procedures for the development of a modern cadastre. Standards are suggested where such standards are grounded in adequate prior experience. However, detailed specifications are provided only through references to other publications. Each of the procedures identified is well established in some locality in North America. The report provides a critical appraisal of the available standards and procedures, organized under the major components of the basic structure of a multipurpose cadastre:

1. A reference frame consisting of a geodetic network;
2. A series of current, accurate large-scale base maps;
3. A cadastral overlay that delineates all cadastral parcels and displays a unique

identifying number assigned to each of them, the latter serving as a common index of all parcel-related land records in information systems; and

4. A series of registers or files that record interests in land parcels, each including a parcel identifier for purposes of information retrieval and linking with information in other land-data files.

1.2.1.1   Densification of the Geodetic Reference Frame

Monuments that are precisely located by geodetic surveys are needed at more closely spaced intervals in most parts of the United States, so that positions of land-related data may be determined. Ideally these monuments should be as densely spaced as the monuments for section and quarter-section corners of the Public Land Survey System (PLSS) (Committee on Integrated Land Data Mapping, 1982).

1.2.1.2   Production and Maintenance of Base Maps

Large-scale base maps locate the major physical features of the landscape at scales of from 1:500 to 1:25,000. The procedures and standards given herein for the production of base maps have been used for the production of single-purpose cadastres but can evolve readily into standards for production of multipurpose cadastres. Their use will assure not only adequate large-scale maps for the purposes intended but also compatibility among the large-scale maps of the separate counties or municipalities.

1.2.1.3   Preparation of Cadastral Overlays

An important task in almost all localities is to establish a legal status for the cadastral overlay as the timely, complete, and available inventory of all existing land parcels. Each cadastral parcel must have a unique identification number. There is need for more than a pictorial representation of parcel boundaries of the kind that traditionally has served limited purposes such as real-property assessment or administration of public services. Functions such as parcel indexing of land-title and other parcel-related records have more demanding requirements for accuracy and especially currency of the maps. However, this does not require that the cadastral overlay actually serve as the legally sufficient statement of property boundaries. For representations that are legally sufficient statements of boundary one normally must refer to recorded surveys that are drafted at a much larger scale or to the legal descriptions in the land-title records.

Ties of property boundary surveys to the geodetic coordinate system are essential; the procedure used to accomplish them will depend on methods available to a local government. Coordinates of property corners can be invaluable for integrating adjacent surveys and will facilitate the eventual automation of the production of maps.

1.2.1.4 Building and Maintaining Land-Parcel Registers and Data Files

The unique parcel identification number provides a means for linking the cadastral parcel to land-data files or registers that contain information about land ownership, use, value, assessment, and other attributes. Other parcel identifiers may be introduced to facilitate the use of these land-information records and thus form a family of parcel identifiers. These are the codes that permit sharing of land-parcel data among the agencies in each jurisdiction and with other jurisdictions that maintain compatible multipurpose cadastres. A system of identifiers may be coded to indicate the political districts to which each parcel belongs.

1.2.2 Suggested Procedures for Linking Other Land Information to the Cadastre

The report describes successful procedures for connecting other files of land-based information to the basic cadastral reference system. Some of these procedures are feasible only where cadastral files and other records are automated.

1.2.2.1 Referencing Other Land Information to the Base Map and Cadastral Overlay

Criteria are offered for judging when maps of the boundaries of land characteristics other than those of parcels are needed. These maps establish overlays distinct from the cadastral overlay. Procedures for describing such boundaries by coordinates, and the potential value of these coordinates for the automation of map production, are described. Limitations of describing the locations of land characteristics by grid cells are compared with those of describing boundaries by coordinates.

1.2.2.2 Direct Comparison of Other Land Information with Cadastral Records

The cadastral parcel can serve as a complete and adequate locational reference for many types of social, economic, physical, and administrative data. Thus it is the building block for statistical comparisons among these land characteristics. However, most natural phenomena, and certain other man-made divisions of land, such as for agricultural production, must be attributed to individual land parcels before direct, statistical comparisons can be made with land-ownership information. To do this precisely requires subdividing all records for each parcel that is crossed by the boundary of significant natural phenomena. Other, more economical, procedures are described for approximating the attribution of natural phenomena to parcels.

1.3   REVIEW OF THE REPORT *NEED FOR A MULTIPURPOSE CADASTRE*

Land-information systems in governments at all levels are characterized in the Committee on Geodesy (1980) report as being either the traditional title or assessment systems or the more recently developed land-planning and -management systems. That report categorized the problems inherent in our present systems as accessibility, duplication, aggregation, confidentiality, and institutional structure.

The concept of a multipurpose cadastre concept was presented as a basis for action to remedy the problems that exist in our current system. The multipurpose cadastre concept was described as "a framework that supports continuous, readily available, and comprehensive land-related information at the parcel level."

1.3.1   Components of a Multipurpose Cadastre

The components of a multipurpose cadastre as described in the earlier report (Committee on Geodesy, 1980) are presented in Section 1.2.1.

That report was concerned primarily with the reference frame, base maps, and cadastral overlay components of the multipurpose cadastre. Other elements were discussed to the extent necessary to provide a complete picture of the system.

1.3.2   Improving Land-Information Systems

The considerable amount of activity aimed at improving land-information systems in the United States and Canada was identified. Programs instituted at the state and county level to improve land-recording procedures, including records indexing, computer data handling, computer mapping of utility information, control densification, and large-scale mapping, have been in progress for a number of years. Several states have undertaken control surveying and base-mapping programs and have developed land-data files. Federal agencies have developed special programs for particular land-information areas, such as the Taxable Property Values Survey by the Bureau of the Census and the Real Estate Settlement Procedures Act studies by the Department of Housing and Urban Development. Federal agencies have funded a number of land-information pilot projects and have assisted local and state agencies in their surveying and mapping activities. An extensive land-information system exists in the Maritime Provinces of Canada, which provides a well-developed model for others to consider.

The variety of participants who will each have a role in the creation of a Land-

Information System was suggested by the statement in the Committee on Geodesy (1980) report that "federal, state, and local governments as well as private contractors have an important role in the development of a multipurpose cadastre." The basic high-order control surveys already are being done by federal agencies; the close-spaced monuments should be set by state or local agencies or their contractors. Small- and medium-scale mapping is well under way by federal agencies. Large-scale maps should be prepared by state and local agencies or their contractors. The basic cadastral surveys of federal land are being performed by federal agencies, while state land is being surveyed by state agencies or their contractors. Local property boundaries should be established by private surveyors with the approval of the chief surveying officer of each county or municipality. The cadastral overlay will be the result of work by surveyors, abstractors, title attorneys, zoning organizations, and courts.

1.3.3 Essential Requirements for a Multipurpose Cadastre

The earlier report (Committee on Geodesy, 1980) listed several essential requirements for development of a multipurpose cadastre. These are as follows:

1. The development of technical standards and specifications and the means to enforce these;
2. The development of linkage mechanisms in order to relate other land information to the basic components;
3. An emphasis on gradual, phased reorganization and quality control of existing governmental functions, rather than creation of new functions and agencies;
4. A focus on the county level as the place where much of the work in developing and maintaining a multipurpose cadastre will occur, with appropriate support by state and federal governments; and
5. The development of qualified personnel through encouragement and support of university research and education.

The report recommended several specific actions. These are as follows:

1. Federal legislation to authorize and support the creation of a multipurpose cadastre in all parts of the nation;
2. Designation by the Office of Management and Budget of a federal lead agency for promotion of the development of multipurpose cadastres;
3. Continued technical studies sponsored by the federal government to identify consistent land information and display standards for use among and within federal agencies and between federal and state governments—these studies should rely on the authority of state governments to adopt the standards and organize the data collection in cooperation with the federal government to ensure compatibility on a

national basis, delegating these functions to county governments where appropriate;

4. Authorization by each state of an Office of Land Information Systems, through legislation where necessary, to implement the multipurpose cadastre.

In summary, the 1980 report described the nature of land-information problems, established the need for a response to those problems, defined the elements and structure of a multipurpose cadastre as a basis for action, identified the development process for a multipurpose cadastre, and recommended specific legislative and administrative initiatives at all levels of government that would lead to development of a multipurpose cadastre.

1.4   MULTIPURPOSE CADASTRE CONCEPTS

1.4.1   Origins

Cadastre is defined in *Webster's Third New International Dictionary* as "an official register of the quantity, value, and ownership of real estate used in apportioning taxes." *Black's Law Dictionary* defines a cadastre as a "tax inventory and assessment of real property."

The origins of what is accepted as the modern cadastre concept are found in the cadastral systems of Continental Europe that were formed during the eighteenth and nineteenth centuries. Like earlier efforts, these were designed fundamentally for taxation or fiscal purposes. Cadastral-system development was associated with a "ground tax" concept wherein most state revenues were obtained by "levying a ground tax, ultimately based on the taxable revenue of the separate ground parcels, and buildings, subdivided according to their different use as agriculture grounds, meadows, orchards, woods, houses, factories, workshops. . ." (Henssen, 1973). The ground tax concept evolved over time into complex differential tax-assessment systems, based in part on differing land uses. These complex systems required supporting land-parcel information arrangements.

It appears that as early as the seventeenth century the Europeans developed an understanding and appreciation of the cadastre concept for purposes beyond taxation. The evolution of the legal or juridical cadastre is traced from this period (Henssen, 1973). The juridical cadastre was conceived as a system for recording and retrieving information concerning the tenure interests in the land that, as with the fiscal cadastre, required the identification of the people holding an interest in the land and the location of those interests. However, the juridical cadastre required a more rigorous delineation of these interests in order to provide for the secure transfer of title to the land.

1.4.2   Evolution of the North American Cadastral Arrangements

The early North American cadastral arrangements were designed to promote quick, efficient, and secure land settlement. The alienation of public or crown lands, as a means of inducing European emigration, was from the outset recognized as a basic function of government in the English colonies. In order to secure the private interests in land, several land-records institutions were established. These were designed to provide legal notice of transactions involving land and included public recordation of deeds, recording acts, abstracts of title, and opinions of title evidence.

Land-survey practices and institutions developed consonant with the desire to achieve and secure alienation of public land. The public-land survey was established during this period.

During the nineteenth century the property tax developed as the primary basis for local government revenues. Institutions and practices associated with the real-property assessment process were firmly established as a function of local government along with the concept of uniform property taxation based on value.

North American cadastral institutions are distinct in their time and place. They differ from those in Continental Europe. However, all share a focus on fiscal and juridical purposes.

The traditional cadastre is often a routine file of parcel-related data designed to meet special purposes with efficiency and timeliness, especially valuation and title. Although these traditional cadastres often draw on data and information from various sources, they are characterized by their special-purpose outputs of products and services. However, the routine use of these files as a source of land information is rarely satisfactory for purposes other than those originally intended. The need is for a multipurpose cadastre designed to *provide* a wide range of relatable land information. The multipurpose cadastre, in form and substance, must rise to the level of a research and comprehensive planning tool, as well as continuing to serve traditional, special-purpose needs.

1.4.3   Nature of a Modern Cadastre

A modern multipurpose cadastre is defined as a record of interests in land, encompassing *both* the nature and extent of these interests. An interest or property right in land may be narrowly construed as a legal right capable of ownership or more broadly interpreted as any uniquely recognized relationship among people with regard to use of the land.

An understanding of the nature and extent of interests in land requires not only traditional fiscal and juridical records but also public-land management, infrastructure, physical, and similar records. These requirements are the inevitable result of the growing complexity associated with a postindustrial society.

The growing complexity of rights and interests in land also requires improvement in the measurement and representation of the spatial extent of rights and interests and in the institutions associated with measurement and representation of land.

Cadastral systems, whether of traditional or modern design, are concerned with information and data about man's division of the land into parcels for purposes of ownership and use. These systems, and the public and private institutions that support them, have developed because of citizen and official needs for information. As relations between people concerning use of the land become more complex, more and better information is needed.

Modern concepts of a cadastral system, as a geographic-information system that employs the proprietary land unit (the cadastral parcel) as the basic reference unit for gathering, storing, and disseminating information, has three basic components. These are as follows:

1. The cadastral parcel, defined as a continuous area of land within which unique, homogeneous interests are recognized. It is defined three dimensionally in recognition of subjacent and superjacent interests and in time.

2. The cadastral record, the source of graphical and/or alphanumeric information concerning both the nature of the interests and the extent of those interests.

3. The parcel index, the system for relating parcels and records.

The cadastral system is a combination of people, technical resources, structure, and organized procedures that results in

1. The official recording of data pertaining to the initial delimitation of cadastral parcels and their subsequent mutation;

2. The official recording of data pertaining to all recognized tenure interests in these parcels;

3. The official recording of other parcel-relatable data; and

4. The subsequent storage, retrieval, dissemination, and use of these data.

1.4.4 The Cadastre as Part of a Larger Geographic-Information System

A cadastre may be regarded as a part of a larger system of land-related information called a Geographic-Information System (GIS). A GIS is any system of spatially referenced information or data. Spatially referenced information or data have a unifying characteristic—association with a specific place on the Earth's surface. A GIS is designed to gather, process, and provide a wide variety of geographically referenced information that may be relevant for research, management decisions, or administrative processes.

The information contained in a GIS may be classified according to whether the focus of the information is on people or on land. Socioeconomic information con-

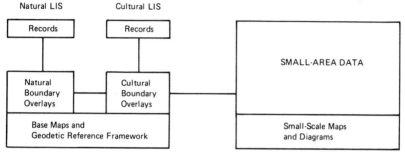

FIGURE 1.1   Types of Geographic-Information Systems (GIS's).

tained, for example, in the files of the Bureau of the Census is people-focused information primarily, although the information has a spatial component. Each element of social or economic data is indexed to a discrete location, as defined in Chapter 6, which may be a municipality or a block or even a building address or parcel number. However, the spatial component is not the principal element around which the system is designed.

If the information in the GIS focuses primarily on the land, then the information is part of a Land-Information System (LIS). This hierarchy of systems is diagrammed in Figure 1.1. The spatial context of the data in a LIS is continuous, as defined in Chapter 6, with locations preferably referenced to a plane coordinate system. The focus in a LIS, the land with its spatial aspect, requires a significantly higher degree of spatial accuracy than that associated with a socioeconomic system. The data content of a LIS is more likely to relate to the physical environment than to social or economic conditions and is normally less subject to questions of personal confidentiality.

LIS's, with their emphasis on land-related information and spatial accuracy, can be divided into two distinct sets. These sets correspond to the cultural and natural divisions of the Earth's surface.

Natural LIS's are concerned with the many ways the Earth is divided according to its physical characteristics such as soils, vegetation, mineral resources, depth to bedrock, and flood hazards. Cultural LIS's are concerned with the divisions of the Earth into parcels by man for purposes of ownership and use.

1.4.5   Distinctive Features of the Multipurpose Cadastre

A multipurpose cadastre is designed to record, store, and provide not only land-tenure and land-valuation information but also a wide variety of parcel-relatable information. It is truly multipurpose in that it not only receives information and data from many sources, but it also provides relatable services and products for many purposes and to many users.

The multipurpose cadastre is the core module of a large-scale, community-oriented information system designed to serve both public and private agencies, and individual citizens, by (1) employing the proprietary land unit (cadastral parcel) as the fundamental unit of spatial organization of land information and (2) employing local government land-record offices as the fundamental unit for information dissemination.

The fundamental importance of individual, decentralized decision making about use of the land by individual citizens or local governments is recognized by use of the proprietary land unit and local government offices. The possibility of greater citizen input to and scrutiny of local systems is enhanced by emphasis on parcels and local offices.

The multipurpose cadastre system is designed to overcome the difficulties associated with traditional, limited approaches by (1) providing in a continuous fashion a comprehensive record of land-related information and (2) presenting this information at the parcel level. The multipurpose cadastre is a public system, operationally and administratively integrated, that supports timely, readily available, and comprehensive land-related information at the parcel level. The multipurpose cadastre concept is built around an accurate spatial framework, base maps, a cadastral overlay tied to legal records of property boundaries, and linkage to land records distributed about many offices and users.

The components of a multipurpose cadastre, as shown in Figure 1.2, are described in Section 1.3.1. Table 1.1 lists some of the many benefits that have been realized

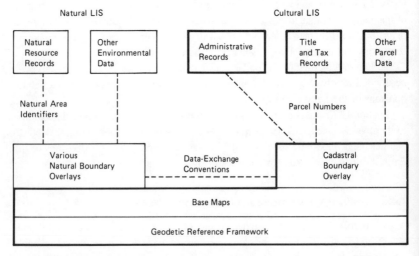

FIGURE 1.2   Components of a multipurpose cadastre (in heavy outline) as the foundation for Land-Information Systems (LIS's).

TABLE 1.1 Some of the Potential Benefits of a Multipurpose Cadastre to Each of the Major Types of Users

Potential Benefits to Local Governments
    Assures that the best available data are used in each public transaction
    Avoids conflicts among land records of different public offices
    Improves accuracy of real-property assessments
    Provides base maps for local planning and preliminary engineering studies
    Provides a standardized data base for neighborhood, municipal, county, or regional development plans
    Avoids costs of maintaining separate map systems and land-data files
    Encourages coordination among public programs affecting land
    Improves public attitudes toward administration of local government programs
Potential Benefits to State Governments
    Provides accurate inventories of natural assets
    Provides accurate locational references for administration of state regulations such as pollution controls
    Accurately locates state ownerships or other interests in land
    Provides a standardized data base for management of public lands
    Provides large-scale base maps for siting studies
    Simplifies coordination among state and local offices
Potential Benefits to the Federal Government
    Provides a flow of standardized data for updating federal maps and statistics, e.g., for the federal censuses
    Provides a data base for monitoring objects of national concern, e.g., agricultural land use and foreign ownership of U.S. real estate
    Provides a reliable record of the locations of federal ownerships or other interests in land
    Provides standardized records for managing federal assistance to local programs such as housing, community development, and historic preservation
Potential Benefits to Private Firms
    Produces accurate inventories of land parcels, available as a public record
    Produces standard, large-scale maps that can be used for planning, engineering, or routing studies
    Speeds administration of public regulations
Potential Benefits to Individuals
    Provides faster access to records affecting individual rights, especially land title
    Clarifies the boundaries of areas restricted by zoning, wetlands restrictions, pollution controls, or other use controls
    Produces accurate maps that can be used for resolving private interests in the land
    Reduces costs of public utilities by replacing present duplicative base-mapping programs
    Improves efficiency of tax-supported government services, as described earlier in this table

in lands that are served by multipurpose cadastres. Further details on how each type of user relates to a cadastral system are given in Section 7.2.

1.5 GOVERNMENT RESPONSIBILITIES

Financial and personnel commitments by county, state, and federal governments are necessary for the creation of multipurpose cadastres. Gradual, phased, incremental establishment is necessary because the legislative and budget processes of governments tend to address short-term, readily identifiable problems rather than long-range improvements. A recent study of the cost of land records in Wisconsin (Larsen *et al.*, 1978) indicates that the greatest expenditures are at the local level. It may be possible to obtain a substantial portion of the cost of development from savings associated with more efficient operation of land-records systems at the local level.

1.5.1 County Government Responsibilities

The content of land records generally is related to functions of local government much more than to functions of state or federal governments. It is at this level of government, close to the individual citizen and to the individual parcel of land, that combinations of human and technical resources with organizing procedures are needed that result in the collection, storage, retrieval, dissemination, and use of land data in a systematic way.

A framework that supports continuous, readily available, and comprehensive land-related information at the parcel level must be local in its nature in order to meet the particular demands of citizens and officials close to the making or implementation of decisions about parcels of land. A local multipurpose cadastre must provide a framework that satisfies the needs of other jurisdictions including state and federal governments and the means to transmit state and federal government land information to the local level.

Some counties for several years have been improving their procedures for recording ownership, indexing, computer data handling, computer mapping, monumentation of Public Land Survey System corners and other points, and standards for accuracy and completeness of their land records.

The result is improvement of individual, routine files of land data. These files are designed to meet special purposes with efficiency and timeliness. However, they seldom are organized so that all the collective data on a given parcel can be available to one local agency. For those counties with adequate resources, modern data-base-management systems can provide a solution.

1.5.2 State Responsibilities

The decisions of state governments, or their lack of decisions, will set the pattern for the development of multipurpose cadastres in each state. The adoption of standards that would assure the compatibility among the individual county and municipal cadastres will depend on the authority of state legislation. Whether a local government will take any initiative at all in organizing a multipurpose cadastre will often be a reflection of the expressed interests of state agencies. Many statewide programs stand to benefit from the establishment of multipurpose cadastres that serve each county, which would provide a wealth of data on needs and resources of the state, plus the geographic framework for referencing state administrative records.

The report of the Committee on Geodesy (1980) recommended that each state create an Office of Land-Information Systems to provide the needed leadership and to administer grants-in-aid to local governments. Without such leadership the development of multipurpose cadastres will remain scattered and of uneven quality.

1.5.3 Federal Guidelines

Guidelines are needed for federally supported programs that may have an impact on cadastral development. Similarly, there are issues related to standards and procedures that must be considered by local officials as they strive to meet local needs for land information. It must be emphasized that, while there are local problems to be addressed by local standards and procedures, nevertheless, Land-Information Systems are dynamic systems and should be compatible with other local multipurpose cadastre systems and with larger networks of information. A balance is needed between demands at the local level and a need to keep open options that permit aggregation of data to higher levels of government.

2
The Geodetic Reference Framework

2.1 BACKGROUND

A geodetic reference framework forms the spatial foundation for the creation of any Land-Information System (LIS). Consisting of monumented points whose locations have been accurately determined with respect to a mathematical framework, this system permits the spatial referencing of all land data to identifiable positions on the Earth's surface. A geodetic reference framework provides not only an accurate and efficient means for positioning data, but it also provides a uniform, effective language for interpreting and disseminating land information.

2.1.1 The National Geodetic Survey

The National Geodetic Survey (NGS) is mandated by law to establish and maintain a National Geodetic Reference System (NGRS) adequate for present and future public needs. The present NGRS is a network of close to 750,000 monumented control points whose precise geographic positions or elevations, or both, have been determined by geodetic surveys. This network is the basic positioning reference for all U.S. mapping and charting, large-scale engineering works, national defense operations, Earth satellite tracking, and a wide variety of other national and local endeavors. It provides the only practical means for determining the relative positions of widely separated points in a common, unified coordinate system. Government and private survey organizations use NGRS stations as absolute reference points for local surveys to minimize the propagation of errors that would otherwise result in the mispositioning of property boundaries, engineering works, and other features.

To eliminate discrepancies between adjacent surveys and to ensure that each land parcel is uniquely positioned and identified, communities have had to abandon local datums and isolated coordinate systems and integrate area systems with the NGRS. This has become especially critical for communities that maintain computerized data bases that cannot tolerate ambiguities.

The NGRS consists of surveys from sources other than the NGS, including other federal agencies, state agencies, and private surveying/engineering firms. Coordination is accomplished by adoption of procedures established by the interagency Federal Geodetic Control Committee (FGCC). By Office of Management and Budget directive and interdepartmental agreement the FGCC reviews governmentwide requirements and programs for geodetic activities and makes determinations if surveys can be practicably and economically improved and permanently monumented to form a part of the national networks. The integrity of the NGRS is maintained by adoption of the FGCC standard "Classification, Standards of Accuracy, and General Specifications of Geodetic Control Surveys." Surveys are forwarded to NGS for adjustment and inclusion in the NGRS by adopting the FGCC standard "Input Formats and Specifications for the NGS Data Base" (Federal Geodetic Control Committee, 1980).

Available from the National Geodetic Information Center (NOAA, NGS, Rockville, Maryland 20852) are maps depicting the available points in the NGRS. Available are horizontal coordinates and descriptions for 250,000 points and elevations and descriptions for another 500,000 points. Availability is in hard copy and digital form (magnetic tape or microform). Recently, the horizontal data base has become available by a direct telecommunications link.

2.1.2   State and Local Geodetic Networks

Responsibility for the coordination of geodetic control activity at the state level varies from essentially no organization or coordination in some states to the existence in other states of strong state geodetic survey agencies. In many states, leadership resides within a state department of transportation, division of surveying and mapping (or equivalent). In 17 states, there is an office of state surveyor (or equivalent), which has been established by state statute to coordinate surveying activity.

The federal government has provided each state in the United States with a plane coordinate system based on the NGRS and one or more zones of either the Transverse Mercator projection or the Lambert Conformal Conic projection. The present State Plane Coordinate Systems were first introduced in the 1930's and are based on the 1927 North American Datum. They have been ratified by legislation in 38 states.

At the regional, county, and local levels, geodetic control activity is carried out both by private surveyors and engineers, in response to the needs of specific projects, and by state and local governments, usually the survey crews of the public works departments.

2.2   GEOMETRIC FRAMEWORK REQUIREMENTS FOR THE CADASTRE

Any Land-Information System requires some method of spatial reference for the data. An adequate geometric framework for such reference must, if it is to serve even the narrowest of purposes of a cadastre, permit identification of land areas by coordinates down to the individual parcel level. The provision of a geometric framework of adequate accuracy and precision to permit system operation at the highly disaggregate parcel level is the most demanding specification possible. It permits ready aggregation of information from the more intensive and detailed level to the more extensive and general level as may be necessary.

The type of geometric framework to be provided for any new land-data system is one of the key determinations affecting the long-term, as well as the initial, utility and efficiency of the system. Any error in this determination should be made on the side of potential long-term utility. A determination to provide a geometric framework more precise and accurate than may be required ultimately will mean that a portion of the capital required to implement the system may be wasted. A determination, however, to provide a geometric framework less precise and accurate than may be required ultimately will mean that most or all of the capital investment required to implement the system will have been wasted. Further, such a misapplied capital investment itself may form an insurmountable impediment to later evolutionary development of the system, since the committed decision will with time make it increasingly difficult and costly to effect any required reforms. In this respect, it is particularly important to resist the temptation to use only paper records of mapped locations as a basis for the development of the land-data system in order to save initial costs.

Because of the importance of the geometric framework for the spatial reference of data to the long-term success of any multipurpose land-data bank, and because that importance is apt to be overlooked by planners and decision makers in their deliberations of other important issues involved in the creation of land-data systems, a brief discussion of certain fundamental concepts that should be applied in the design of the geometric framework for any land-data bank system is in order.

2.2.1   Fundamental Concepts

An effective multipurpose land-record system must be able to store in machine-readable form a wealth of data essential to sound land-use planning and management. Historically, such data have been typically stored on maps. Consequently, certain concepts that apply to the design and preparation of good maps also apply to the design and implementation of the geometric framework for a land-data system.

Any accurate mapping project requires the establishment of a system of survey control. This survey control consists of a framework of points whose horizontal and vertical positions and interrelationships have been accurately established by field surveys and to which the map details are adjusted and against which such details can be checked. The survey control system should be carefully designed to fit the specific needs of the particular map being created. For multipurpose application, it is essential that this survey control system meet two basic criteria if the maps are to be effective planning and management tools. First, it must permit correlation of real-property boundary-line data with topographic, earth-science, and other land and land-related data. Second, it must be permanently monumented on the ground so that lines on the maps may be reproduced in the field when land-use development and management projects reach the regulatory or construction stage. That is, the survey control system must support the production of finished maps, the points and lines of which not only accurately reflect both cadastral and earth-science field conditions but also can be readily and accurately reproduced on the ground. This capability is important not only to the use of the maps but also to their maintenance in a current condition.

Conceptually, the geometric framework for a Land-Information System is the equivalent of the survey control system for a map; and the same principles apply to its design and implementation.

2.2.2 Design Issues

In the design and development of a geometric framework for a land-data system, three important issues require resolution. One issue relates to the type of mathematical map projection to be used as a basis for the geometric framework. The second relates to the density—or control-station spacing—requirements for the framework. The third relates to the accuracy requirements for the framework. The determinations concerning these three issues should be made in light of the basic concepts set forth above.

The curved surface of the Earth cannot be represented on a plane surface such as a map without distortion that increases with the size of the area involved. Accordingly, a projection that can be used to transform geodetic positions on the surface of the Earth to corresponding plane coordinate positions on the map is required. Properly selected, the map projection becomes a powerful mathematical tool for performing rigorous survey computations as well as providing a basis for the accurate graphical representation of the mapped data.

A number of projections have been used as a basis for the preparation of large-scale maps. The federal government through the NGS has provided practical projections for use in local surveying and large-scale mapping operations through the State Plane Coordinate Systems. The principles underlying the State Plane Coordinate

Systems and the data necessary for their application in the development of a geometric framework for a land-data system are set forth in publications of the NGS (Special Publication Numbers 235 and 62-4).

The universality of coordinate values expressed in terms of the State Plane Coordinate Systems provides a compelling argument for the use of these systems in the development of a geometric framework for land-data systems. The state plane coordinates can be transformed readily, precisely, and within known accuracy limitations into other coordinate systems, thereby permitting the correlation and use of the data in regional, state, and national as well as in the local systems.

We recommend the use of the State Plane Coordinate Systems as the basis for the recording of positions in local land-data systems in the United States. Selection of any other projection should be done reluctantly and only after most careful consideration.

The second major issue to be addressed in the design of a geometric framework for a land-data system is the necessary density of the horizontal control network. If the positional integration of the land data is to be accomplished solely by graphic means—the necessary correlations being provided solely by reference to the coordinate grid shown on the maps—only the density of control needed for the maps is required. If, however, the integration of the positional information is to be accomplished numerically, relatively high-density standards are required. Numeric integration of the data should be an essential feature of any modern land-data system, and the density and accuracy requirements of the horizontal survey control should be determined accordingly.

Monumented points of known position on the State Plane Coordinate System should be so distributed throughout the area concerned as to permit their ready use in the collection of both cadastral and earth-science data. *Typical recommendations range from 0.2 to 0.5 mile (0.3 to 0.8 km) between monuments in urban areas to 1 to 2 miles (1.6 to 3.2 km) in rural areas* (Ziemann, 1976; McLaughlin, 1977). *We concur with these recommended densities of monumented points.* In those areas of the United States covered by the PLSS, monuments established at approximately one-half-mile intervals at section and quarter-section corners and at the centers of sections would meet the system design for control stations. An accurate position of the center point of a section is required in order to provide a proper basis for the compilation of cadastral maps and data.

Ideally, the entire area concerned should be covered at a uniform density with a simultaneously adjusted network of control survey stations. As a practical matter, however, the necessary survey work will have to be carried out over an extended period of time. To provide the required uniformity in such successive surveys, a higher-order control net may have to be established. The spacing of the higher-order stations can be up to 10 miles (16 km) but is usually 3 to 5 miles (5 to 8 km). In any case, the local survey network should be an integral part of regional, state, and national control nets.

With respect to accuracy, the determining factor will be the extent to which the control survey stations are to serve multiple purposes. Similar to the above, if the integration of the positional data is to be done graphically, a relatively low order of accuracy will be required for the horizontal control network, such as that attendant to the federal classification of third-order, class II (Federal Geodetic Control Committee, 1978). If, however, the data are to be integrated numerically and if the control surveys are to have multiple applications, minimum accuracies at least attendant to the federal classification of third-order, class I, or second-order, class II, should be met.

As a final consideration, all the control stations should be monumented with substantial, stable, readily recognizable survey monuments. The monuments should be accurately tied to at least three reference marks, and documentation should permit their ready recovery and use both for the maintenance and extension of the basic control net and, importantly, in the collection and use of the land data themselves.

2.2.3 The Public Land Survey System

The Public Land Survey System (PLSS) was organized starting in 1785 as the geometric framework for all lands that at that time were the responsibility of the federal government and part of the public domain. Those lands, and the vast areas that subsequently were acquired by the U.S. Government, today exist within 30 states that contain 80 percent of the land of the nation and are home to 56 percent of its population.

The PLSS provides a unique identifier for each mile-square section in these lands and may be used to identify fractions of sections as small as 2½ acres. The boundary lines are defined by monuments placed at intervals of ½ mile in the original survey of each section commissioned by the federal government. If these monuments are lost, then replacements are valid only if placed in a survey that "follows in the footsteps of the original surveyor." A summary of the history and structure of the PLSS is presented in Chapter 2 of the report of the Committee on Integrated Land Data Mapping (1982).

The property boundaries defined by the original PLSS monuments have the attributes of registered property boundaries in that they are immune to relocation by "adverse possession," even by fence lines that are long established. The correct, legal locations (or relocations) of the monuments are the very foundation that holds the PLSS together and without which the definitions of property boundaries in vast areas of the United States would come unraveled.

For all the nonfederal lands in the United States that are subdivided according to the Public Land Survey System (PLSS), we recommend that the geodetic reference framework for the cadastre be the section corners and quarter-section corners of the PLSS, including the center point of each section. In the non-PLSS states, an even distribution of selected property corners or eccentric corners and right-of-way monuments should form an equivalent framework. Each county (or municipal) cadastre program should be made responsible for assuring that these points have been relocated and monumented according to the legally established procedures prevailing in each state and properly connected to the National Geodetic Reference Framework to obtain geodetic coordinates.

The spacing of survey control points at intervals of $1/2$ mile that is recommended here is consistent with the standard for urban areas recommended in the preceding section. For rural areas in the PLSS states, this relatively close spacing is justified as a completion of the monumentation of the basic framework of rural property lines, none of which would be delineated if only the section corners at 1-mile intervals were accurately located. At least one end of every PLSS property line depends on reference to a quarter-section corner to establish its position, and every PLSS property line would be left floating if only the section corners at 1-mile intervals were located, except in the relatively few areas of the PLSS where no quarter-section corner monuments ever were put in place. Figure A.5 of Appendix A.1 shows how the basic framework of property lines is completely defined by the survey control points at half-mile intervals.

This recommended density of survey control in the 30 PLSS states and the 20 "metes and bounds" states has the following important advantages:

1. It provides a common, consistent, and accurate system of control for both real-property boundary lines and topographic mapping. In the 30 PLSS states the boundaries of the original government land subdivision form the basis for all subsequent property divisions and boundaries; thus the accurate re-establishment of the U.S. Public Land Survey quarter-section lines and corners permits the compilation of real-property boundary-line maps and supports the compilation by usual photogrammetric methods of topographic maps. Moreover, in any state that requires all new land subdivision plats to be tied by surveys of a specified accuracy to the available geodetic control points, these boundary-line maps can be readily and accurately updated and extended into newly developing areas. The inclusion of State Plane Coordinates on these plats can be readily required by local subdivision regulations.

The system permits the accurate correlation of property boundary-line information with topographic details supplied by aerial mapping. This placing of property-boundary and topographic data on a common datum is essential to sound mapping for the ultimate development of a multipurpose cadastre, as well as for planning and engineering purposes, yet such a common control datum is rarely used. The establishment of State Plane Coordinates for a dense network of survey control points permits the transfer of details supplied by aerial mapping, including hypsography, to property boundary-line maps by simple overlay methods. Savings in office research time made possible during the planning and design phases of municipal public works projects by having all available information—topography, property boundaries, and control—

accurately correlated on one map are great. Moreover, such complete and correlated information and control makes possible the consideration and analysis of many alternate routes for such public works facilities as trunk sewers, water transmission lines, and major trafficways and of many alternative solutions to sewerage, drainage, water supply, and transportation problems.

2. It provides a practical horizontal control network readily usable by both private and public surveyors and engineers for all subsequent survey work within the urban area. The control system outlined places a monumented, recoverable, control station of known position and known elevation at half-mile intervals throughout the area mapped. This monumented control net not only expedites such engineering surveys as are made almost daily, year in and year out, by such public works agencies as city engineering and water departments; county and state highway departments; and sewerage, transportation, airport, and harbor commissions for planning, design, and construction layout purposes, but also correlates and coordinates all their survey work throughout the entire area mapped. In this regard the control system outlined is particularly valuable in providing a common system of control for the precise location and mapping of underground utilities, both public and private.

3. It makes the State Plane Coordinate System available as a practical matter for property boundary-survey control, thus providing the means for using State Plane Coordinates in boundary descriptions supplemented in 30 states by the PLSS. In most of the United States it would be the first meaningful control net available to the land surveyor. Property corners, in many urban areas, have long been inadequately monumented and, therefore, readily susceptible to loss. Points of beginning in a metes and bounds legal description have often depended on unmonumented corners or on street and highway intersections that cannot be relocated precisely. The accurate retracement of property boundaries under such conditions is extremely difficult and costly, and the accurate mapping of such boundaries by public agencies is well nigh impossible. Moreover, the uncertainties of title and accompanying litigation resulting from such conditions become more and more unsatisfactory as urbanization intensifies and land values increase. By requiring the relocation and permanent monumentation of the Public Land Survey corners, which cover 80 percent of the United States, the recommended system will do much in itself to stabilize real-property boundaries and make the control net of great value to private land surveyors. By utilizing this control, local land surveyors can, without changing their methods of operation or incurring any additional expense, "automatically" tie all their surveys to the State Plane Coordinate System, and all bearings used in land surveys, plats, and legal descriptions will be directly referenced to grid north and thereby to geodetic north. If the use of the State Plane Coordinate System is to be encouraged, it is essential that it be made available in this manner to the local land surveyors. While the State Plane Coordinate System was devised by the U.S. Coast and Geodetic Survey (now the National Geodetic Survey) in the early 1930's, it has seen only limited use in many areas by land surveyors and local engineers, who have generally been unfamiliar with both

the system and the methods necessary to carry geodetic control down to the property being surveyed. Only by making state plane coordinates available to the surveyor through the PLSS, which he understands and employs constantly, can their use in land and local public works engineering surveys be achieved practically.

4. It would reinforce the many other values of the PLSS in the 30 states now using that system. Lines of the PLSS drawn on the cadastral maps are guides for reserving land for future public use, taking land for public use, describing districts within which public regulations are to be applied, or locating and aligning proposed public works projects. The PLSS is economically adaptable to the latest survey techniques. The PLSS supports the eventual creation and maintenance of a multipurpose cadastre and a multipurpose land-data bank.

The cost of control surveys for aerial mapping projects typically accounts for one quarter to one third of the total cost of the finished maps. Such control surveys are unusable by local engineers and surveyors, and their costs are largely unrecoverable. By allocating to the control survey work a relatively small additional amount of the total resources that might be available for mapping, far more effective and useful finished maps can be obtained, and a valuable and permanently useful system of survey control can be provided concurrently. The only significant increases in cost actually assignable to the control system proposed are relatively small and are solely those incurred for the relocation and monumentation of the land-survey corners and the small amount of additional traversing required to coordinate these corners. These additions to cost amount to approximately 20 percent of the total cost of an urban mapping project—small in relation to the benefits.

2.3   OUTLOOK FOR NEW TECHNOLOGY

To date, a geometric framework that would support a multipurpose cadastre exists in a small fraction of the counties in the United States. Indeed, in only about 10 percent of the 500 counties designated by the U.S. Department of Commerce as "leading" counties in terms of economic activity is there in place an existing primary geodetic framework of sufficient density (spacing of 3 to 5 miles or less) (or 5 to 8 km or less) to serve as a logical starting point for densification to a level that would support the cadastre. Although there are multitudes of PLSS section-corner monuments in 30 states, relatively few of them have been located with reference to a coordinate system in a manner that will support planning and engineering. Because of limitations of staffing, the National Geodetic Survey has had to turn down or table requests from scores of counties for establishment of primary nets. Thus, most counties are not even in a position to begin the establishment of the local geometric framework required to support a cadastre.

Even where the primary geodetic net is already in place, it could well take a decade for a single surveying party to complete a project of densification sufficient to support a multipurpose cadastre for a typical county (Brown, 1977). Moreover, total costs for a county of average size could easily entail several million dollars, as described in Section 7.3.2. Examples of such long-term county programs are described in Appendix A.

Fortunately, much progress has been made toward the introduction of more effective and efficient positioning technologies that permit substantially lower per-unit costs for large-scale densification of survey control. Some of these, such as photogrammetric triangulation, are well-established technologies. Others, such as the Global Positioning System (GPS), are just becoming available.

2.3.1 Photogrammetric Triangulation

The most mature and well-documented alternative to field surveying is photogrammetric triangulation. With this technology an area is covered by a block of photographs having a high degree of overlap (at least 60 percent in both directions) so that each point of interest is imaged in nine or more photographs. A small number of previously surveyed points serve as a framework of control within which all other points are established by an interpolative process based on measurements of the coordinates of images on film. The most rigorous and accurate approach to photogrammetric triangulation is by the "bundle-adjustment method" (Brown, 1973). This involves the simultaneous least-squares adjustment of the sets (or bundles) of rays from all photographs to all measured ground points in a process that also simultaneously recovers the projective parameters (position and attitude) of all photos. In a more advanced version referred to as the "bundle adjustment with self-calibration" the process is expanded to include the estimation of additional unknown parameters (or error coefficients) that describe residual systematic errors in the observations. In blocks with relatively sparse control, the bundle adjustment method with self-calibration can provide a significant improvement in accuracy.

One especially important feature of the bundle-adjustment method is that it does not require that horizontal control be distributed fairly uniformly throughout the area in the manner considered desirable for densification by conventional ground methods. Rather, it suffices if such control is distributed about the periphery of the block with a spacing corresponding to that of about every fifth photograph. With such control, accuracies in planimetry turn out to be nearly uniform throughout the photogrammetric block, no matter how many photographs are contained in the block. This means that densification by the bundle method requires far less primary control than is needed for densification by ground methods. As is pointed out in Brown (1971), savings resulting from sharply reduced requirements for primary control may often be sufficient to pay for all costs of photogrammetric densification except those associated with monumentation. In a subsequent study, Brown (1977) found that photogrammetric triangulation could provide as much as a 3-to-1 cost advantage over first-order ground traversing. Furthermore, at least in principle, almost any desired accuracy can be produced by photogrammetric triangulation (it being primarily a function of the scale of the photography). According to recent tests conducted by NGS

(Lucas, 1978) accuracies approaching 1/100,000 of the flying height can be approached when highly exacting procedures are employed (including adjustment by the bundle method) and when a modern aerial mapping camera of 6-inch (150-mm) focal length is used in conjunction with a focal-plane reseau (a device that projects at uniform intervals throughout the format sharply defined points of reference that serve to account for effects of film deformation). Accuracies of better than 1/50,000 of the flying height have been reported by Europeans employing cameras without focal-plane reseaus. In other words, accuracies of 0.1 ft (3 cm) are being realized when the flying height is 5000 ft (1500 m) or less.

However, aside from a few pilot tests of relatively limited scope, no significant application of photogrammetry to the establishment of a geographic data base has been reported to date. The technology reached maturity over 5 years ago, and its capabilities and advantages have been widely disseminated throughout the technical community. It appears that a major impediment to its practical utilization stems from a lack of recognized standards for photogrammetric triangulation. Many methods exist, and their accuracies may vary by a factor of 10 (Brown, 1973). This proliferation of methodology is confusing to the uninitiated and will be a continuing obstacle until suitable photogrammetric standards are adopted.

2.3.2   Inertial Surveying

Another important emerging technology is inertial surveying. This technology had its beginnings at the outset of the 1970's in a military system designed originally for artillery surveying to accuracies of ± 30 ft (±9 m) over open traverses of about 120 miles (193 km). However, the original Position and Azimuth Determining System (PADS) displayed a potential for considerably greater accuracies, and subsequent improvements in hardware, software, and operating procedures have rapidly elevated the system to a level of accuracy competitive with conventional second-order traversing.

The heart of an inertial surveying system is an inertial navigational unit designed for use in aircraft. The Litton LN-15 inertial navigator, employed in the first of the inertial surveying systems to be developed, has a basic drift rate in application to aircraft navigation of about 1 km (3300 ft) per hour of flight. On the other hand, in applications to ground surveying this performance is effectively improved by a factor of several thousand. Such drastic improvement has come about largely through exploitation of the fact that in ground surveying the vehicle bearing the inertial unit can be brought to a dead stop at frequent intervals. During such stops, which are called *zero-velocity updates*, residual components of velocity attributable to drifts in the system will be sensed (these would be zero in a flawless system). Such discrepancies generate observational equations that are processed in real time by an onboard computer. This leads to estimation and partial compensation for drifts in the inertial unit. Further compensation can be made through apportionment of errors of closure on points whose coordinates are known (control points). An analysis of results from a large number of field tests provides the following estimates of the accuracies of an inertial surveying system (Mancini, 1977):

> Elevation: 10 + 8 cm/h
> (0.33 + 0.26 ft/h)

> Horizontal position (lat. or long.): 13 + 12 cm/h
> (0.43 + 0.39 ft/h)

These estimates are based on operations involving zero-velocity updates (stops of about 30-sec duration) at intervals of 3 to 5 min and an apportionment of errors of closure on control points. An interesting facet of an inertial survey is that, up to a point, the faster it can be performed the more accurate it is, for the buildup of error is predominantly a function of time rather than distance. For this reason a helicopter is generally the preferred vehicle whenever feasible.

Accuracies of inertial surveying for relative positioning for successive points that are relatively close together (as in urban densification) are dependent mainly on the constant term (or "threshold") in the expressions given by Mancini (1977), because the time required to travel from one point to another is only a matter of a few minutes. Thus, accuracies in relative horizontal positioning of such points can be expected to be roughly on the order of 13 to 18 cm (0.4 to 0.6 ft). For points separated by 800 m (about 0.5 mile) this corresponds to a proportional accuracy of about 1 part in 4400. On the other hand, relative horizontal positions for a pair of points, say 10 min and 10 km (6 miles) apart, could be established to a proportional accuracy of almost 1 part in 50,000. This indicates that under certain circumstances inertial surveying can produce accuracies equivalent to those of conventional second-order traversing. However, it appears that an improvement by about a factor of 2 to 3 is needed if positions of closely spaced urban monuments are to be established to the desirable level of accuracy of about 5 cm (0.2 ft). With the rapid developments in this field, it is probable that this accuracy will be achieved in the near future.

The Bureau of Land Management, U.S. Department of the Interior, has had extensive experience with inertial surveying technology for cadastral purposes. They have employed a system consisting of six components, all interconnected by a system of electrical cables and installed in a helicopter. These components are the inertial-measurement unit, data-recording unit, power-supply unit, data-processing unit, control and display unit, and auxiliary battery. Total weight mounted in a helicopter without battery is about 125 kg (275 lb). The control and display unit is mounted on the helicopter instrument panel. The remaining components are secured to a plywood pallet mounted in the rear seat area. An optical hoversight device is mounted in front of the pilot, which enables him to hover directly above any point on the ground.

The system has been used to determine geographic coordinates and elevation of PLSS section corners in relation to other points having known coordinates and elevation. As the system is transported by helicopter, incremental velocity changes are automatically recorded in three directions at short intervals of time. These velocities are integrated to obtain a continuously updated set of coordinates and elevation of the instantaneous location of the system. The coordinates and elevation of all other section corners in the township are determined by hovering over each point and reading the system display directly. In actual field use, the point coordinates are stored in the computer memory for later computer processing.

The cost of currently available inertial surveying units is quite high—on the order of several hundred thousand dollars. Should more compact and cheaper inertial surveying units eventually become almost as widely used as first-order theodolites are today, their use could extend to routine property surveying.

2.3.3 Satellite Doppler Positioning

Both photogrammetric and inertial surveying provide efficient means for densification of control within an existing primary geodetic network, the former requiring considerably sparser control networks than the latter. However, as mentioned earlier, in numerous communities the requisite network of primary control is not in place. This obviously precludes the start of densification until an appropriate primary net is established. Unfortunately, the NGS has the staff and resources to establish only a few primary nets each year. It appears, then, that here again new technology must be brought to bear if an expeditious resolution of the problem is to be realized. In the near term, the best prospect for meeting this need appears to be by means of satellite Doppler positioning using signals from Navy Navigational Satellites.

Accuracies obtainable from Doppler positioning using the most advanced observational procedures coupled with the most exacting methods of data reduction appear to be reliable on the order of 10 to 20 cm (0.33 to 0.66 ft) for relative positioning of stations separated by up to 100 km (about 60 miles), observing from 40 to 60 satellite passes in common with each other or in common with a continuously occupied base station located in the general vicinity (Brown, 1979). For separations beyond 100 km, accuracies deteriorate slowly with increasing distance to the level of about 30 to 40 cm (1.0 to 1.3 ft) for separations of 400 km (250 miles) (and observations of 40 to 60 passes). At mid-latitudes about four days of observations would generally be required to observe 40 to 60 satellite passes. Currently it appears that accuracies can be improved to the 5- to 10-cm (0.16- to 0.33-ft) level for separations under 100 km, as a result of a combination of factors, such as exercise of stronger local tracking configurations (e.g., use of two or more base stations), refinements in data reduction, deployment of the new series of NOVA satellites, use of improved local oscillators, and improvements in Doppler receivers.

A scenario for the establishment of a primary network of, say, 18 stations within a typical countywide area by satellite Doppler positioning to accuracies on the order of 10 cm (0.33 ft) might proceed as follows (such a net would be sufficient to support a project of densification by photogrammetry). A pair of base stations would be established at convenient existing points of the primary net (these need not be in the county itself but preferably would be near opposite extremities of the county). In addition, two mobile units would be deployed, each occupying a designated station for a period of 4 to 5 days before being moved to another station. Such a field campaign could be completed comfortably within a month. All observations gathered would be subjected to a simultaneous least-squares adjustment leading not only to the determination of positions of the stations but also to refined values for orbital parameters as well as estimates of coefficients of error models associated with each station for each pass (in essence a process of self-calibration). Proportional accuracies in relative positions for stations separated by 10 km (about 16 miles) could be expected to be on the order of 1/100,000; for stations separated by 50 km (about 30 miles) proportional accuracies would be on the order of 1/500,000. The cost of such a survey could be expected to be about $75,000 (or roughly $4000 per station) of which about $50,000 would be for field work and $25,000 would be for data reduction. Data reduction could be expected to be completed within 60 days after the conclusion of the field campaign.

From the foregoing it appears that in the task of establishing primary geodetic nets satellite Doppler positioning has reached the point of competitiveness with conventional first-order surveying in terms of cost, accuracy, and timeliness. While other satellite methods in the developmental stage promise further improvements on all three counts, for the next 5 years or so satellite Doppler positioning is likely to remain the primary alternative to conventional surveying for establishing local geodetic nets suitable for densification by other means.

2.3.4 Global Positioning System

The advanced technology of the NAVSTAR Global Positioning System (GPS), now being developed by the Department of Defense (DOD) and with the participation of the National Geodetic Survey, the U.S. Geological Survey, and the National Aeronautics and Space Administration (NASA) in the development of a geodetic receiver, could provide tremendous benefits to surveyors. This proposed cost-effective system has the potential capability of providing relative positional information at the 1- to 2-cm (0.03- to 0.07-ft) level with a few hours of observing time.

The GPS will be a successor to the Navy Transit satellite navigation system. The present system of positioning with Doppler receivers uses signals from the Transit satellites. Both systems were designed primarily to obtain almost instantaneous position determination for purposes of accurate, worldwide, all-weather navigation. Geodetic accuracies can be obtained, however, by using special receivers for the GPS signals and special observation and data-processing techniques. However, the technology described here is new, and the expectations for its use in land surveying may change rapidly.

The GPS, which should be fully operational in 1987, will contain a minimum of 18 satellites grouped in six orbital planes of three satellites each. The first GPS satellite was launched in 1978; five are currently in orbit. A minimum of five satellites will be maintained until initial implementation of the operational system in 1984. Recent experimental tests indicate that the signals from GPS satellites can be used to determine relative positions with 1- to 2-cm accuracies in essentially all-weather conditions in a matter of a few hours over distances of 100 km or less. Over distances of 1 km or less, accuracies of 5 mm have been obtained in 1 or 2 h (Counselman, 1982). These capabilities have important implications for upgrading, densifying, monitoring, and maintaining geodetic control networks.

If the development of GPS technology lives up to present expectations, it could shortly revolutionize surveying and supercede all current horizontal positioning methods. It would be fast, inexpensive, and accurate. Points could be located wherever needed. They would not have to be located on mountaintops, and no observing towers would be required because the control points would not need to be intervisible. The receivers would be small, lightweight, and easily portable as backpacks. They could be set up, turned on, and left to receive and record signals for later processing at a central site. Observing times would be on the order of an hour or two, day or night, in almost any weather. Furthermore, unlike classical methods, which do not determine positions and elevations at the same time, GPS is a three-dimensional system (Counselman and Steinbrecher, 1982; MacDoran *et al.*, 1982).

With the cited advantages of GPS, it appears that this system can increase productivity, reduce costs drastically, and produce accuracies that are not attainable by any other means. Equipment that could be used for survey control for the average county has been on the market for only a short time. At present, a receiver costs about $100,000, although the price can be expected to decrease as the equipment comes into wider use.

2.3.5 Other Positioning Technologies

Other positioning technologies should be mentioned. For example, the advent and continuing improvement in electronic theodolite instrumentation is probably the greatest advancement in surveying instruments since the development of electronic distance-measuring (EDM) instruments. Several models of electronic theodolites are available with various accuracies, applications, sizes, and prices. Although the concept originated and the first model appeared in the late 1960's, the greatest acceptance, demonstrated by volume of sales, has been in the last 5 years. Improvements since 1970 have reduced volume and weight, increased accuracy, and provided greater versatility.

Electronic theodolite instrumentation integrates an electronic theodolite, with EDM and the automatic recording of measurement data. Provision for connection to any on-line computer is often included, as is the internal computing capability. The distinguishing feature is that the theodolite is electronic. The essential components such as uprights, telescope, axes, clamp, and motions are largely identical with those of an optical theodolite, but the electronic theodolite has electronic circle readings.

These universal instruments excel in most surveying applications, including reading coordinates, using program control and data memory. Whether for control surveys, topographic surveys, or engineering surveys, speed and accuracy of measurements make electronic theodolite instrumentation a cost-effective solution for the volume of required survey data. This data-capture process is the beginning of the automated data flow.

The NASA Goddard Space Flight Center is currently developing a multibeam Airborne Laser Ranging System (ALRS), which ranges simultaneously to six ground-based passive reflectors with centimeter precision. By flying over a target grid at two altitudes, the system can provide a snapshot of the target positions (latitude, longitude, and altitude) over an extended area, which is limited only by the range and maximum altitude of the aircraft. High-altitude ($\sim$18,000-m) research aircraft, such as the U-2 or RB-57, can potentially survey areas up to 60,000 sq km in one 6-h flight with error growth rates on the order of 1 cm per 100 km of baseline from the reference origin. In general, the error growth rate per unit baseline varies inversely with the maximum aircraft altitude.

The approach being developed by NASA is to invert the usual configuration of the laser-ranging system originally designed for ranging to satellites by placing the ranging and pointing hardware in an aircraft and replacing the expensive ground stations by low-cost ($<$\$1000) passive retroreflectors. The instrument would be constructed on a standard aircraft pallet so that it can be easily removed and reinstalled. This capability eliminates the need for a dedicated aircraft and allows special flights to be scheduled quickly in response to increased seismic activity.

The system is necessarily multibeam since the location of the aircraft is not known with centimeter precision at each point where a set of range measurements is made. Thus, a minimum of four simultaneous range measurements is required—three to resolve the new coordinates of the aircraft and one to acquire information on the relative locations of the ground targets. The ALRS for geophysics applications will be capable of ranging simultaneously to six retroreflectors. At a laser repetition rate of 10 pulses per second (pps), a potential 1.3 million individual range measurements can be made during one 6-h flight of a high-altitude aircraft. Computer simulations have demonstrated that, with range biases and single-shot rms standard deviations on the order of 1 cm, the ALRS will be capable of resolving baseline distances on the order of 100 km to the subcentimeter level from such a platform. Furthermore, the data-reduction technique simultaneously resolves the aircraft position to the centimeter level at each point in the flight path where a laser pulse is transmitted.

The system is expected to be a powerful new research tool for monitoring regional crustal deformation and tectonic plate motion because it will provide a snapshot of the target positions (all three axes) over an extended area with high spatial resolution. In addition to its geophysical applications, such an instrument would clearly allow the rapid verification of existing ground-survey networks and permit further densification on a regional scale with target spacings on the order of 5 to 20 km (Degnan, 1982).

There is another laser system that has been prepared for very precise geodesy. This system, referred to as the Spaceborne Geodynamics Ranging System (SGRS), would employ a single satellite in a circular orbit of 100-km altitude and 50° inclination (Smith, 1978). Aboard the satellite would be a laser-ranging system that would be directed in rapid succession toward up to several hundred ground-based retroreflectors distributed over regions of interest. Computer simulations indicate that accuracies in relative positioning on the order of 1 cm (0.03 ft) or better can be expected from a 6-day mission for stations separated by up to 300 km (about 200 miles). The report from the Workshop on the Spaceborne Geodynamics Ranging System (1979) recommended that study of the approach be continued with the goal of implementing an actual demonstration in conjunction with the Space Shuttle at the "earliest possible date."

2.3.6   Conclusions Regarding Feasibility

The technological developments described above do not change the fundamental concepts and principles laid out in this report. Rather, their impact will be on the costs of the various alternative procedures for building a multipurpose cadastre. Their effect will be to increase the feasibility of a cadastral records system.

Because of economic considerations, most of the satellite systems considered would be applicable mainly to the establishment of primary geodetic nets and would not displace densification by photogrammetric, inertial, or conventional traverse methods. A possible exception would be a system having capabilities comparable with those projected for the GPS. If the system could successfully operate amid the obstacles of an urban environment, it would suffice to have a single base station at a convenient point in each county operating in conjunction with any number of mobile units operated by private surveyors performing routine surveys. Alternatively, in difficult areas one could envision a GPS system used in conjunction with a compact (second- or third-generation) inertial system, the former providing nearby temporary control for the latter.

The foregoing considerations make it clear that emerging technology will be of increasing, and ultimately dominant, importance in the establishment of the multipurpose cadastre. A rational program for widespread implementation of the cadastre concept must accord due weight to such developments. Some jurisdictions need to improve their system of ground control now. They cannot wait for the new technology. The costs of survey-related projects in the short term and the uncertain pace of technological development make investment in improved ground control by traditional methods a wise one. Jurisdictions that will have significant growth in the near future and jurisdictions with particularly poor systems of existing control are candidates for investment now. Whether investment occurs now or in the future, however, there should be adherence to the basic principles of a geodetic reference framework—large-scale base maps and a cadastral overlay.

3
Base Maps

A base map is the graphic representation at a specified scale of selected fundamental map information; used as a framework upon which additional data of a specialized nature may be compiled (American Society of Photogrammetry, 1980). Within the multipurpose cadastre, the base map provides a primary medium by which the locations of cadastral parcels can be related to the geodetic reference framework; to major natural and man-made features such as bodies of water, roads, buildings, and fences; and to municipal and political boundaries. The base map also provides the means by which all land-related information may be related graphically to cadastral parcels.

Good planning and engineering practice dictate the preparation of large-scale maps as a basis for sound community development and redevelopment. In urban areas, and particularly in growing urban areas, such large-scale maps are currently being compiled at an unprecedented rate by photogrammetric methods. Relatively simple changes in the specifications governing these photogrammetric mapping operations can make the resulting maps not only more effective planning and engineering tools but can, at relatively little additional cost, lay the foundation for the eventual creation of a multipurpose cadastre.

3.1 ALTERNATIVE FORMS OF MAPS

There are three fundamental forms that may be used to represent map information: (1) line map, (2) photographic or orthophotographic map, and (3) digital map. The conventional line map is a line and symbol representation of natural and selected man-made features on a coordinate reference system. Different line, symbol, and area colors are used to aid in distinguishing between water features, man-made objects, wooded areas, and contours. A line map is produced from scribed, inked, or pasted-on line copy. A photographic map is a photograph or assembly of photographs on which descriptive cartographic data, marginal information, and a coordinate reference system have been overprinted. The photographs may be uncontrolled, nominally vertical aerial photographs, or they may be rectified photographs, with image displacements due to camera tilt removed. An orthophotographic map is similar to a photographic map with the exception that, in generating the orthophotographs from conventional aerial photographs, image displacements caused by both camera tilt and terrain relief are removed. Photographic images on an orthophotographic map are therefore in their correct orthographic map position. Digital maps have evolved in recent years with the development of powerful data-processing systems that have made it possible to collect and store digitized map data. Manipulation and merging of the digitized data and selective retrieval of desired levels of map information, either in graphic form as a plot or a printout or in numerical form as a body of data, make the digitized representation of map information (virtual map) a very flexible form (Thompson, 1979).

Each of the forms of map information (line map, photographic map, and digital map) has its advantages and disadvantages as candidates for a base-mapping medium in a multipurpose cadastre. The majority of map-producing organizations today are producing line maps. To a considerable extent, a line map can selectively control the type and amount of information to be shown on the base map. However, line maps are the most difficult and expensive to update in a timely fashion. Photographic maps can be readily updated with the collection and processing of new photography, and they contain a large amount of terrain surface detail. As a base map, however, the photographic map may have more detail than desired, without the possibility for control of the type and amount of information to be shown. Image displacements in the photographic map due to camera tilt and terrain relief are removed in the orthophotographic map, with the additional expense of the differential rectification process necessary to produce the orthophotograph. With the development of the techniques of automated cartography, digital mapping promises to be the form most responsive to the requirements for flexible selection of type and amount of base-map information and for regular base-map updating. Only in this form can map information that has been collected at different scales and in different formats be efficiently merged, digitally, and displayed together. Necessary digital mapping standards will evolve as this new technology matures in future production mapping environments. Their development is being advanced currently by a Committee on Digital Cartographic Data Standards organized in 1982 by the American Congress on Surveying and Mapping, with the sponsorship of the U.S. Geological Survey (Moellering, 1982).

3.2 SOURCE MATERIAL

There are three alternatives to be considered when evaluating the source materials to be used for base maps in a multipurpose cadastre: (1) existing maps (line, photographic, or digital), (2) existing maps updated with new map information during the course of the cadastre operation, and (3) new maps. The tradeoffs among the alternatives are map uniformity and accuracy versus the cost of new mapping.

Unless there has been a consolidation and standardization of the mapping effort within the various departments to a single, unified mapping activity, and a recent large-scale mapping program completed, existing maps for a given county or municipality are likely to be incomplete, out of date, or otherwise less than ideal for use as a mapping base for the cadastral overlay. The basic mapping functions in a typical local government environment, at the county level for example, are generally spread among a number of departments or divisions, primarily (1) assessment, (2) public works, and (3) planning. The base-map requirements for each of these departments vary, especially with regard to map scale, format, and content. This situation fosters a general lack of coordination among departments, duplication of effort, and often an absence of adequate, professional mapping personnel (Archer, 1980). The accuracy of the existing maps may be unknown or not adequate for present-day urban requirements. The cost of immediate new mapping, on the other hand, may appear to be prohibitive. Thus the need to consider the alternatives.

Substantial savings can be realized by adapting a new map system to existing base maps, if they are adequate. For example, in Prince William County, Virginia, a new Mapping Division was created as a consolidation of the mapping efforts of the Finance and Assessments Division, the Public Works Department, and the Planning Office. The division developed a unified mapping program using 1:2400-scale base maps, with 320 maps covering the 345 square miles of the county. Prior to consolidation, the property identification maps alone numbered 597. These maps originally included 197 at a scale of 1:4800 and 400 at a scale of 1:1200. In preparation for the 1:2400 base-map scale, the property identification maps are being photomechanically reduced or enlarged and then digitized so that computer plotted overlays can be drawn at the base-map scale and any other scale desired. A cost analysis of the Prince William County system indicates that $158,000 was being expended annually, prior to consolidation of the mapping program, to maintain the separate mapping efforts. With an anticipated increase in the number of larger-scale property identification maps, this $158,000 was expected to increase to $178,000. It was estimated that complete coverage of the county at the 1:2400 base-map scale would cost $350,000, which reduces to about $1000 per square mile. However, since the Public Works Department's topographic maps at the 1:2400 scale were found to be of high quality, cost of completion of the base mapping was only $150,000, or $435 per square mile. With the significantly reduced number of maps, the annual maintenance cost for the unified mapping program is $75,000, less than half of the cost before program consolidation (Archer, 1980).

The state of Missouri is taking the approach of a new, statewide comprehensive mapping program. The State Tax Commission supported a study in late 1979 by GRW Consulting Engineers, Inc., of Lexington, Kentucky, to assist in the design and development of a new statewide base-mapping program. The state has a land area of approximately 69,000 square miles, with 2,300,000 land parcels administered in 115 assessment jurisdictions. Recommendations for mapping bases included photographic maps using aerial photographs, orthophotographic maps, and planimetric line maps. Rectified photographic maps would be used at 1:2400 and 1:4800 scales where relief was not excessive. Orthophotographic maps would be used at 1:2400 and 1:4800 scales where topography was excessive and for all maps at a 1:1200 scale. The planimetric line map would be used where 1:600-scale base maps are required. Of the 115 assessment jurisdictions, 109 counties covering an area of 66,437 square miles required the full mapping program. The six jurisdictions excluded were those in the Kansas City and St. Louis areas, containing over half of the state population.

The base-map needs and costs projected in 1980 by the consulting engineers for

TABLE 3.1 Projected Coverage and Costs of the Missouri Property Mapping System (109 of 115 Assessment Jurisdictions)

Land area, in square miles		66,437
Total map sheets (32″ × 34″)		26,636
1:4800-scale rectified photographs	2,276	
1:4800-scale orthophotographs	16,264	
1:2400-scale rectified photographs	22	
1:2400-scale orthophotographs	1,871	
1:1200-scale orthophotographs	5,936	
1:600-scale planimetric maps	287	
Total cost estimate		$11,631,128
aerial photography	$ 889,428	
control analytics	2,557,465	
intermediates	5,225,914	
base-map sheet master	65,182	
cadastral control	1,355,970	
final base-map sheet	1,537,169	
Average per county		
Land area, in square miles		610
average map sheets (32″ × 34″)		244
average total cost		$ 106,707
average cost per map sheet		$ 437
average cost per acre		$ 0.27
average cost per parcel		$ 7.88
average cost per square mile for this statewide program		$ 175

this new statewide program appear in Table 3.1. In the initial stages of this program, which are just recently under way, orthophotography is being used as the standard base-map system, at a cost that is ranging between $250 and $500 per square mile.

3.3 CONTENT

Design of the base-mapping data content and structure must be flexible enough to allow a variety of users to relate the cadastral parcels to specific types of base information. This objective can readily be achieved by creating and maintaining the base-mapping data in a coordinated series of different levels or overlays. Photographic and orthophotographic base maps at a minimum contain the complete photographic image of the terrain surface covered, to which other levels or overlays may be added to create the complete base map.

The primary base-map datum is the geodetic reference framework used to establish the location of all other features. The following reference systems are in current use throughout the United States:

1. Geographic Coordinates (latitude and longitude)
2. Universal Transverse Mercator (UTM) rectangular coordinates
3. State Plane Coordinates

Geographic coordinates provide the principal system used for computation of geodetic control-point positions. The UTM rectangular coordinate system is a metric worldwide system of predominate use in federal mapping environments. State Plane Coordinates are most commonly used at the state and local levels, currently defined in English units but with metric units also widely available with completion of the National Geodetic Survey readjustment of the North American Horizontal Datum in 1983. Because of the greater familiarity with their use at the local level, State Plane Coordinates are normally used as the geodetic reference framework in current implementation projects and are recommended for local multipurpose cadastres (see Section 2.2.2).

Natural and cultural features that are relatable to a cadastral parcel form the next most important levels of base-map data. One of these levels includes all streets, roads, railroads, and airports, with their associated names. Another level includes all permanent buildings and other structures greater than a specified size. A third level includes all water features such as perennial and intermittent streams, natural and man-made lakes and ponds, reservoirs, canals, and aqueducts and their associated names. A fourth level includes boundaries of civil (governmental) jurisdictions at all levels: state, county, city, and township. Other secondary levels of natural and cultural features, such as contours, floodplains, wetlands, vegetation cover, land use, and utility lines, may be included selectively in the base-map composite.

Use of a number of different levels or overlays of base-map data is essential to provide flexibility in meeting the different requirements of different map users. Drafted overlays must be precisely registered in position to each other as illustrated in Figure 3.1 (Archer, 1980). A planner may desire, for example, to use as a working base a composite of the land-use, floodplain, and base-map overlays. The greatest flexibility in base-map content, to satisfy user requirements, is in digital mapping. Map information can be separated digitally into a maximum number of data levels, updated most efficiently, and plotted precisely on a single base sheet using any specified number of data levels as required. At the same time, standards and procedures must yet be established to control the level of map content detail as map scale is changed over wide ranges.

Overall, the base map that supports a multipurpose cadastre must provide as a minimum enough planimetric detail for locating ownership boundaries referenced to natural features, such as stream and lake shorelines, or to man-made features not as yet tied to the coordinate system, such as highways and railroads. Desirably, it should

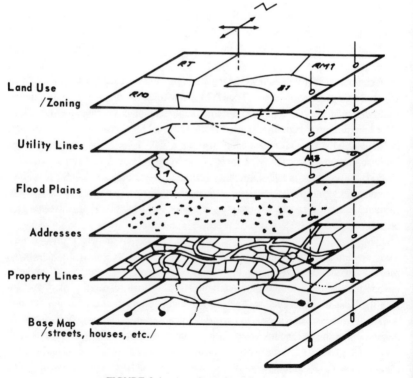

Land Use
/Zoning

Utility Lines

Flood Plains

Addresses

Property Lines

Base Map
/streets, houses, etc./

FIGURE 3.1   A registered overlay system.

show all objects related to the location of real-property boundaries, such as fences or driveways, at reasonably frequent intervals.

3.4 ACCURACY

Accuracy of the horizontal and vertical position information on the base map is fundamentally a function of the map scale and contour interval, respectively. National Map Accuracy Standards (Appendix B) have long been used as the primary standard to control the accuracy of plotted map information. For scales larger than 1:20,000, which include essentially all base maps that would be used to support a cadastral overlay, standards for horizontal accuracy specify that 90 percent of the points tested shall be plotted on the map within 1/30 inch of their true position. Standards for vertical accuracy specify that 90 percent of the points tested shall be shown in elevation within one half of the contour interval used on the map.

The Photogrammetry for Highways Committee of the American Society of Photogrammetry has prepared specifications for large-scale mapping for highways, with a horizontal accuracy requirement that 90 percent of all planimetric features be plotted within 1/40 inch of their true position (U.S. Department of Transportation, 1968). This is a more stringent requirement than the comparable 1/30 inch required by National Map Accuracy Standards and has also been suggested by the Task Committee for Photogrammetric Standards of the American Society of Photogrammetry in their recently proposed Accuracy Specifications for Large Scale Line Maps. Either the 1/30-inch or 1/40-inch requirements have been adopted by nearly all users in their base-mapping specifications for large-scale property-ownership maps.

The requirement that the base map of a local record system be compiled according to National Map Accuracy Standards (Appendix B) is primarily due to the need for the base map to satisfy the engineering needs of public works departments. When accurate information is necessary, specific boundary lengths would come from a recorded plat, boundary description, or other report of survey, not from scaling the cadastral overlay on the base map. A new Engineering Map Accuracy Standard has been proposed by the Committee on Cartographic Surveying of the Surveying and Mapping Division of the American Society of Civil Engineers (ASCE). These standards are intended to provide a clearer communication of accuracy requirements between those having the need for the map and those preparing the map. Also included are specific field-testing procedures to assess the compliance of the map with the standards.

Whatever standard is agreed on, the need exists for quality control within a base-mapping program to verify conformance of the mapping to standards. An accuracy-check ground survey is necessary to determine the ground positions of checkpoints for comparison with their corresponding mapped position. Evaluation of the checkpoint results, using the National Map Accuracy Standard, can be accomplished by first computing the standard error, $\sigma = [\Sigma x_i^2/(n - 1)]^{1/2}$, where $x_1, x_2, \ldots x_n$ are the errors observed at the n checkpoints. Assuming a normal distribution for the checkpoint errors, the 90 percent error can then be computed as 1.645σ and then compared with the accuracy standard. Care must be taken to ensure that the map position of checkpoints has not been deliberately altered to accommodate standard map symbolization. The proposed ASCE Engineering Map Accuracy Standards specify the use of 20 well-defined and widely distributed points for determination of the checkpoint discrepancies. At least 15 percent of the checkpoints shall appear in each quadrant of the map. The accuracy-check ground survey shall be at least of an order of accuracy equal to that of the control survey on which the map is compiled. This proposed standard would not identify specific error values (such as 1/30 inch or one-half contour interval as cited in the National Map Accuracy Standards) but rather leaves those values for negotiation between the engineering client and the cartographic engineer. Evaluation of the accuracy-check ground survey is made by computing limiting standard errors in each of the X, Y, and Z positions, for comparison with the values specified during negotiation, and also computing limiting mean absolute errors in each of the X, Y, and Z positions, again for comparison with the values specified (American Society of Civil Engineers, in preparation).

The scale of the cadastral map system is principally a function of the size of the predominant land parcel. This criterion generally corresponds to the level of land value or degree of urbanization. Listed in Table 3.2 are the scales that have been selected almost universally for each type of area. In the Missouri Tax Mapping Project, for example, the scale of the base maps varied with lot frontage, indicated in the second column in this table.

Contours may only need to be included on the base map for specific users with a requirement for topographic detail. The added expense is substantial. Contour interval would be selected in conjunction with the map scale, the terrain relief, and the elevation information requirements. Typical combinations are listed in Table 3.3.

TABLE 3.2   Suggested Base-Map Scales

Type of Area	Customary Lot Frontage	Comparable Base-Map Scale	Metric-Map Scales
Urban	15' to 40'	1:600 (1″ = 50′)	1:500
Urban	50' to 90'	1:1200 (1″ = 100′)	1:1000
Suburban	100' to 180'	1:2400 (1″ = 200′)	1:2000, 1:2500
Rural	200' and greater	1:4800 (1″ = 400′)	1:2000, 1:5000
Resources		1:12,000, 1:24,000	1:10,000, 1:25,000

TABLE 3.3 Appropriate Contour Intervals for Suggested Map Scales

Customary		Metric	
Base-Map Scale	Typical Contour Interval	Base-Map Scale	Typical Contour Interval
1:600 (1″ = 50′)	1′, 2′	1:500	0.5 m
1:1200 (1″ = 100′)	1′, 2′, 5′	1:1000	0.5 m, 1 m
1:2400 (1″ = 200′)	2′, 5′	1:2000	0.5 m, 1 m, 2 m
1:4800 (1″ = 400′)	2′, 5′, 10′	1:5000	0.5 m, 1 m, 2 m
1:12000 (1″ = 1000′)	5′, 10′, 20′	1:10000	1 m, 2 m, 5 m
1:24000 (1″ = 2000′)	5′, 10′, 20′, 40′	1:25000	2 m, 5 m, 10 m

3.5 OUTLOOK FOR NEW TECHNOLOGY

3.5.1 High-Altitude Photography

Typical mapping cameras with nominal 6-inch focal-length lenses are flown at altitudes below 25,000 ft above mean terrain for the collection of aerial photography. With the increasing usefulness of orthophotographic maps, interest has grown in high-altitude photography using cameras with focal lengths of 6, 12, 24, and 36 inches. The higher the altitude, with the same focal-length camera, the smaller the relief displacement that must be corrected in the photograph for a given amount of terrain relief on the ground. Availability of modified commercial jet aircraft with pressurized cabins has increased the operational flight altitude to 50,000 ft above mean sea level.

Since 1978, the U.S. Geological Survey (USGS) has been developing a National High-Altitude Photography program consisting of both black-and-white panchromatic and color infrared 9-inch × 9-inch photographs taken at a flight altitude of 40,000 ft above mean sea level. The black-and-white photographs are taken with an aerial camera with a focal length of 6 inches, resulting in a photo scale of 1:80,000, each frame representing nearly 130 square miles on the ground. The color infrared photographs are taken by a camera with a focal length of 8.25 inches, resulting in a photo scale of 1:58,000, each frame covering nearly 68 square miles. Standard enlargements are 2×, 3×, and 4×. With flight lines running in a north-south direction, the black-and-white camera exposes a photograph over the center of each USGS 7½-minute mapping quadrangle. The color infrared photographs are particularly useful in resource inventories, agricultural monitoring, and pollution detection.

The NASA U-2 aircraft, based at the NASA Ames Research Center, have a maximum operating altitude of from 65,000 to 70,000 ft above mean sea level. The ground resolution of U-2 imagery collected at 65,000 ft above mean terrain

TABLE 3.4 Alternative Camera Configurations for High-Altitude Photography

Designation	Lens	Film Format (in.)	Ground Coverage at 65,000 ft	Nominal Resolution at 65,000 ft
Vinten (four)	1¾ in. f.l. f/2.8	7 mm (2¼ × 2³/₁₆)	25.9 km × 25.9 km (14 n.mi × 14 n.mi.) (each)	10–20 m
1²S Multispectral (four bands) K-22	100 mm f.l. f/2.8	9 × 9 (4 at 3.5)	16.7 km × 16.7 km (9 n.mi. × 9 n.mi.)	6–10 m
RC-10	6 in., f/4	9 × 9	29.7 km × 29.7 km (16 n.mi. × 16 n.mi.)	3–8 m
RC-10	12 in., f/4	9 × 9	14.8 km × 14.8 km (8 n.mi. × 8 n.mi.)	1.5–4 m
HR-732	24 in., f/8	9 × 18	7.4 km × 14.8 km (4 n.mi. × 8 n.mi.)	0.6–3 m
HR-73B-1	36 in., f/10	18 × 18	9.8 km × 9.8 km (5.3 n.mi. × 5.3 n.mi.)	0.5–2 m
Itek Panoramic (optical bar)	24 in., f/3.5	4.5 × 50	3.7 km × 68.6 km (2 n.mi. × 37 n.mi.)	0.3–2 m
Research Camera System (RCS)	24 in., f/3.5	2¼ × 30	2 km × 20 km (1.1 n.mi × 11 n.mi.) usable	0.1–1 m

varies from 0.1 to 20 m (0.3 to 65 ft) as shown in the tabulation of camera configurations in Table 3.4 (National Aeronautics and Space Administration, 1978). While being used primarily in water-resource and land-use management studies, and in providing ground-truth support for satellite imagery investigations (e.g., Landsat), high-altitude photography has definite potential as a data source for base-map information both as a photograph map and in digital form. Two NASA U-2 aircraft are available on a cost-reimbursable basis for collection of high-altitude photography. Over one third of the United States already has U-2 photographic coverage available, with primary concentrations of coverage over the eastern and western regions of the country.

3.5.2 Satellite Systems

With the launching in July 1972 of the Earth Resources Technology Satellite (ERTS-1), later renamed Landsat 1, concentration by NASA on gathering information from the Earth's surface began, nearly 15 years after Sputnik 1 in 1957. The success that Landsat has enjoyed is generally attributed to its long life and its repeated coverage of the same regions, rather than to its ability to produce images of high resolution. Landsat 1 was followed by Landsat 2 in 1975 and Landsat 3 in 1978. The technical characteristics of these satellite systems, and others, are presented in Table 3.5 (American Society of Photogrammetry, 1980). The low resolution of Landsat— 1:1,000,000 and 1:500,000 scales with 79- and 40-m-square ground-resolution picture elements or "pixels" (that is, about 260 or 130 ft square)—precludes the recording of small cultural features. The positional accuracy of well-defined points on a multispectral scanner (MSS) frame has been reported by Colvocoresses (1975) to be as good as 50 m (164 ft). Landsat images have typically been enlarged by some users by four times, to a scale of 1:250,000. The USGS has compiled and printed black-and-white and color image maps from Landsat frames, at scales ranging from 1:250,000 to 1:1,000,000. Map and chart revision has been accomplished successfully on selected features such as bodies of water, vegetation, and bold cultural objects with scales in some cases as large as 1:50,000. A study in Prince Georges County, Maryland, used the Landsat image to determine the area of various census tracts. Table 3.6 has the comparison of the Landsat results with the Metropolitan Council of Governments' (COG) values. While the direct application of Landsat imagery to support cadastral base mapping is limited, Landsat data can be interpreted to differentiate among a broad variety of surface features. This information can be put to practical use in such applications as agricultural crop forecasting, rangeland and forest management, mineral and petroleum exploration, land-use management, water-quality evaluation, and disaster assessment. Map information from these applications can ultimately be merged with the cadastral base map and overlay to relate the information to the land parcel. Use of Landsat imagery is important because of the large volume of frames available and the continuance of the project with Landsat D. While future

TABLE 3.5 Satellites of Photogrammetric Importance[a]

Satellite	Sensor	Spectral Band	Altitude (km)	Nominal Scale	Ground Resolution (m)	Width of Cover (km)	World Land Coverage	Operational Period
Landsat 1 and 2	Multispectral scanner	0.5–1.1 μm, 4 bands		1:1,000,000	79	185	±82° N–S; almost complete negligible	1972–1978
	Return-beam vidicon	0.47–0.83 μm, 3 bands	919					1975→
Skylab	Multispectral camera S190A	0.4–0.7 μm		1:2,860,000	30[d]	150		
	Photographic camera S190B	0.4–0.88 μm	435	1:948,500	20[d]	108	Negligible	1973
Landsat 3	Multispectral scanner	0.5–1.26 μm, 5 bands		1:1,000,000	79			
	Return-beam vidicon	0.5–0.75 μm, broadband	919	1:500,000	40	185	By request	1978→
Seasat	Synthetic-aperture radar	Microwave L-band	794–	1:500,000	25	100	72° N–S;	1978→

System	Sensor	Spectral band	Altitude (km)	Scale	Resolution		Coverage	Operational
LSA Spacelab	Metric camera RMK 30/23	0.1–0.3 μm	250					
	Synthetic-aperture radar	Microwave, 6 band			30	9	negligible	
Landsat D	Thematic mapper MSS	0.45–12.5 μm	705		30		By request	1982→
	Multispectral scanner	0.5–1.1 μm			79			
SPOT[c]	Multispectral scanner	0.5–0.9 μm	822	1:760,000	10	60	Global eventually	1984→
Shuttle	Large-format camera	0.4–0.9 μm	227–417	1:912,000	≥10[d]	208 × 417	By request	1983
Shuttle	Shuttle imaging radar-A	L-band	278	1:500,000	40 × 40	≥50	By request	Late 1981

[a] Reproduced with permission from *Manual of Photogrammetry*, 4th ed., p. 896. Copyright 1980 by the American Society of Photogrammetry.
[b] European Space Agency.
[c] SPOT (System *P*robatoire d'*O*bservation de la *T*erre) French system.
[d] Pertains to film cameras where ground resolution pertains to one line pair in the image plane. Other values in this column are pixel-size or ground-sample distance.

TABLE 3.6   Comparison of Land-Area Estimates for Sample Census Tracts within the Prince Georges County Study Area

Census Tract Number	Area of the Tract by Metropolitan COG (acres)	Area of the Tract Measured on Interactive System (Landsat) (acres)	Difference in Tract Area Measurement (COG-Landsat) acres	percent
(A) 8011.02	4035	3961	74	1.8
(B) 8012.01	4502	4102	401	9.8
(C) 8012.05	3224	2854	370	8.1
(D) 8019.03	1609	1577	32	2.1
(E) 8022.01	1287	1366	−79	−5.7
(F) 8028.01	2009	2058	−49	−2.3

satellite systems are scheduled to carry sensors with a reported 10-m ground resolution, only after the resulting materials have been evaluated will their increased usefulness to cadastral base mapping be known.

3.5.3   Digital Mapping and Interactive Graphics

The advantages resulting from the collection, storage, and manipulation of base-map information in digital form have been presented earlier. The flexibility of being able to assemble a composite map of different levels of digital map data, and update and extract those levels in a timely manner as new information becomes available, is only possible with an interactive graphics system working with the map information in digital form.

Digital data acquisition may be from previously existing maps, from new mapping, from photogrammetric stereomodels, from ground surveys, or from other terrain information data sources. The usual method of data collection from maps is by manually following the map feature lines on a digitizer table, a tedious operation with a high probability of errors, caused by either duplicating or omitting information. Automatic line-following instruments, usually with the assistance of an operator, improve the accuracy of the data collection. A second approach to automatic data acquisition from maps is to use a scanning device with either a single-element detector or a linear array. Working with the separate overlays (levels) of the map information, this procedure requires a significant amount of computer time to identify and connect together the individual segments of lines in the required "vector" format.

Direct digitizing from photogrammetric stereomodels is facilitated by the use of linear or rotary encoders on the axes of the photogrammetric plotting instruments. Either stereomodel coordinates or photographic image coordinates are recorded initially, with computer program processing transforming the positions into the ground reference coordinate system. The direct link between the stereocompiler and the digital storage eliminates the usual intermediate manuscript preparation stage and subsequent digitizing from the manuscript. This methodology yields a reduction in human intervention and results in ultimately higher accuracies on the digital map (Delaware Valley Regional Planning Commission, 1980). Elevation data can be formatted either as contour lines, profiles with elevations recorded at regular intervals or at breaks in terrain slope, or as geomorphic points along drainage lines or hillcrests, for example. Some photogrammetric instruments are equipped with automatic image correlators that produce a high density of elevation data points.

Each level of map information is stored in its own "layer," in conjunction with other like elements, in the manner depicted in Figure 3.1. This allows for retrieval of any desired combination of levels, such as roads and contours or buildings and property lines. The layering also provides the greater flexibility in producing maps and in an easier update process.

It is essential that the cadastral parcel layer of a digital map system contain complete topological references. That is, each property-line segment in the cadastral overlay must have its own unique identifier and a record that includes the identifiers of its end points as well as the parcels that it bounds. Each end point of property-line segments also must have a unique identifier and a record identifying the line segments and the parcels that meet there. Attached to these point and line records should be other information on their locations, the date and accuracy of the location measurements, and how the points can be relocated in the field. Only with such complete topological and survey data can a digital cadastral overlay be a "living map" that is readily updated as conditions change and that submits readily to automated tests for its completeness and consistency. The same is true for any other overlay intended to be a complete and up-to-date public record, e.g., of a utility system.

The initial interactive graphics task is usually data editing. While data editing may be done by producing a graphic plot on a digitally controlled plotting table, the typical procedure would be to display the data on a digitally controlled cathode-ray tube (CRT). The operator can then determine overlapping, erroneous, or missing data and either make the corrections on the CRT or request a redigitization of the manuscript or stereomodel.

Raw digital data are rarely in form for immediate use, and extensive computer preprocessing is typically required to arrange the data in an appropriate format. This preprocessing may include coordinate transformation from stereomodel coordinates to the ground reference system such as Universal Transverse Mercator Coordinates or State Plane Coordinates. Most data-acquisition schemes acquire far more data than are actually required in the final data files. Therefore, techniques of data compression must be employed to reduce the amount of data to a manageable quantity. Another requirement of the preprocessing system is format conversion, to convert elevation data, for example, interchangeably between contours, profiles, and regular grids of recorded elevation.

Storage of the enormous amounts of digital map data requires an organized system for retrieval. For most modern systems, magnetic tape is the basic storage medium. Header information on each tape will identify the map area, the type of information, and the format in which it is recorded. If digital map data are to be transferable from one facility to another, format standards must be established and used. One such standard has been prepared for data exchange between graphic data bases (American Public Works Administration Research Foundation, 1979b).

Digital map data are being collected in increasingly massive volumes. The Defense Mapping Agency has digitized the contour data on the 1:250,000-scale maps of the entire United States. These data have been turned over to the USGS for storage maintenance and dissemination to users. The USGS has a long-range objective of producing a digital cartographic data base that will contain essentially all of the information now shown on the existing 1:24,000-scale topographic quadrangle maps, both elevation information and planimetric data. This data base will initially contain 11 types of base-map data as follows:

1. Reference Systems—geographic and other coordinate systems except the Public Land Survey System.

2. Hypsography—contours, elevations, and slopes.

3. Hydrography—streams and rivers, lakes and ponds, wetlands, reservoirs, and shorelines.

4. Surface Cover—woodland, orchards, and vineyards.

5. Nonvegetative Features—lava rock, playas, dunes, slide rock, and barren waste areas.

6. Boundaries—political jurisdictions, national parks and forests, and military reservations.

7. Transportation Systems—roads, railroads, trails, canals, pipelines, transmission lines, bridges, and tunnels.

8. Other Significant Man-Made Structures—buildings, airports, and dams.

9. Geodetic Control, Survey Monuments, Landmark Structures.

10. Geographic Names.

11. Orthophotographic Imagery.

Other national mapping organizations in Canada, Great Britain, and Australia are also producing digital cartographic data bases.

Within the private sector, digital data bases are being collected in a number of industries, notably those in petroleum, mining, timber, and public utilities. Four companies with extensive digital data bases across the United States are Phillips Petroleum Company, Tobin Research, Inc., Petroleum Information, Inc., and Stratigraphic Services Company. Much of the data base collected includes digitized section corners of the Public Land Survey System. The digital maps created are exploration maps and lease and ownership maps. These vary in scales from 1:24,000 to 1:1200.

The interactive graphics system is the working tool for digital data storage and manipulation. The hardware and software elements work together in the following functions:

1. Creation of a digital map data base including textual, numeric, and graphic information on geographic facilities and statistical data;
2. Editing of a digital map data base;
3. Selective retrieval of various subsets of data from the data base for presentation on graphic displays and alphanumeric displays and for making digital maps;
4. Producing data tapes in a standard interchange format containing selected subsets of the digital data base;
5. Producing reports from various subsets of data in tabular form.

A procurement specification for an interactive graphics system has been prepared under the Computer Assisted Mapping and Records Activity System (CAMRAS) program by the American Public Works Administration Research Foundation (1979a).

Implementation of the digital-mapping capabilities described above has expanded enormously over the past 10 years, with hundreds of systems currently in place. The

TABLE 3.7   Comparison among Approaches to Developing a Digital Map System

Considerations	Creating System from Scratch	Buying Some Software	Buying Turnkey Software System	Buying Turnkey Hardware/ Software	Buying GIS Services
Dependence on supplier	Very low	Low	High	Very high	Nearly complete
Time until system functions	Long	Long-moderate	Little	Very little	Not a problem
Initial costs	Low	Moderate	Moderate	High	High
Labor costs (user)	High	Lower	Moderate	Moderate	Very low
Risk/uncertainty	High	Lower	Low	Low	Low
Customizing	Complete	Complete	Moderate	Moderate	Varies
Required user technical skill	Extremely high	High	Moderate	Moderate	Quite low
Use of existing resources	High	High	Moderate	Low	Very low

typical system consists of a mainframe minicomputer, disk and magnetic tape units, station interface hardware, graphics software, graphics work stations, and a plotter. Other optional peripherals such as line printers, hard-copy units, alphanumeric terminals, and on-line stereoplotters may also be a part of the system. Acquisition of the interactive graphics capability may occur in one of a number of ways, from the complete development of the entire hardware/software components to the purchase of a "turnkey" hardware/software system to the purchase of the capability as a service. The advantages and disadvantages of the possible approaches are presented in Table 3.7 (Dangermond and Smith, 1980).

4

Cadastral Survey Requirements and the Cadastral Overlay

A basic component of the multipurpose cadastre is a cadastral overlay delimiting the current status of property ownership. The individual building block for the overlay is the cadastral parcel, an unambiguously defined unit of land within which unique property interests are recognized. The overlay will consist of a series of maps showing the size, shape, and location of all cadastral parcels within a given jurisdiction.

Cartographically, the cadastral maps should be viewed as overlays to the large-scale base maps. However, the term "overlay" does not imply that the relationship between the cadastral maps and the base maps is defined by the coincidence of their graphic plots. Rather, the cadastral boundaries are lines connecting points that have unique identities and records, through which they may be located on the ground. Accurate placement of these points on the cadastral overlay does not improve the accuracy of the definition of the boundary, which must be documented elsewhere. The purpose of accurate plotting is simply to make the maps themselves more useful and easier to maintain.

There are several legal mechanisms for establishing cadastral overlay standards, mechanisms that also impart a legal and institutional stability to the cadastral overlay. These mechanisms include (1) land subdivision laws that require surveying, mapping, and recordation according to prescribed standards; (2) regulation of legal parcel descriptions and associated field work by statute as well as by common law; (3) recordation of as-built plans; (4) administrative regulations issued by the courts that register property boundaries in certain states (notably Massachusetts and Hawaii); (5) designation of integrated survey areas; and (6) greater use of the official map in connection with the master plan. The official map is a long-used device (Beuscher and Wright, 1969) that is described in Appendix A.1.

In counties where the delineations of property boundaries by field surveys must be approved by a public office, as in the state of California, it may be possible for the cadastral overlay (including the supporting numerical records based on field surveys) to be tied directly to the legal documents that define the property boundaries, as they are in the cadastres of Continental Europe. This type of public register of property boundaries now occurs in two counties in Massachusetts where registration of both title and boundary includes about half of all parcels.

4.1   CREATION AND MAINTENANCE OF CADASTRAL OVERLAYS

The development of a cadastral overlay will consist of a series of integrated operations, entailing the compilation of land-tenure information and the publication of cadastral maps. Ideally, an area will be chosen for implementing the cadastral-mapping program within which the geodetic reference framework and the large-scale base-mapping program have been established.

There may be a temptation to initiate the cadastral overlay program before an adequate base map is available. The cadastral map is a highly valuable, tangible product that can garner public recognition and support for a multipurpose cadastre initiative. Furthermore, the land-tenure overlays are an invaluable tool in developing other aspects of the cadastral program. However, such a shortcut to a cadastral map should be discouraged, for experience has shown the following:

(a) It is often difficult subsequently to transfer the parcel information to the higher-quality, uniform large-scale mapping base; and

(b) There is always the danger that the product may be misused with a resulting loss in consumer confidence (McLaughlin, 1975).

Another important prerequisite for a cadastral mapping program is a budget for continuing maintenance of the maps. The front-end investments in improved cadastral maps will be wasted if insufficient resources are allocated for keeping them constantly up to date. The annual cost of the technical personnel needed for this work may average more than 5 percent of the original cost of the maps. However, most of the counties that have invested in the elaborate type of digital map system described in Section 3.5.3 have found their costs of map maintenance to be substantially reduced.

We recommend that the updating of cadastral overlays be scheduled so as to assure that they will reliably show any new or changed land parcels that have been in existence for two weeks or more. Where the overlays are used by the recorder of deeds to display the parcel numbers used for indexing the land-title records, this updating should occur within one week.

4.1.1 Comprehensive versus Iterative Mapping Programs

The land-tenure information initially obtained for the development of the cadastral-mapping program will undoubtedly be of mixed quality. Recorded subdivision plans may be available for some areas. For others perhaps only rudimentary deed descriptions will be available. There are two very different approaches to overcoming this difficulty:

(a) Development of a comprehensive survey program designed to establish the bounds of all parcels in advance of the cadastral mapping program and

(b) Development of an iterative mapping program based initially on the existing information base but improved over a period of time as higher-quality information becomes available.

A comprehensive survey program designed to establish the bounds of all cadastral parcels within a jurisdiction has its attractions. The information provided by such a program would presumably be of the highest possible quality with respect to both spatial accuracy and the depiction of all appropriate parcel evidence (particularly possessory evidence in jurisdictions where legally recognized). However, the total economic and human resources, together with the time required to accomplish such a monumental undertaking, often will prove prohibitive. The range of costs may be anywhere from $5 to $50 per parcel or more, depending on such factors as (1) the sizes of the parcels, (2) the quality of the base map, (3) the quality of previous local surveys and their records, (4) whether property corners must be located with typical accuracies of 1–2 ft in rural areas or 0.1 ft in urban areas, and (5) the proportion of costs being assigned to the cadastral overlay.

The only realistic course of action may be to implement the cadastral overlay program in an iterative manner, initially using existing information resources. The minimum requirement for this process is that all parcels must be accounted for, and there must be a capability of correlating the overlay to the base maps.

To support the continuing improvement of local cadastral survey records, we reiterate the recommendations in Section 3.5.4 of the report of the Committee on Geodesy (1980) that

1. Lawyers and surveyors promote state legislation that would make the recording of survey plans for conveyance or subdivision mandatory; all new deeds be based on a reliable survey, similar to those required by the plat laws or section-corner filing acts that exist in some states; and the American Congress on Surveying and Mapping and the American Society of Civil Engineers propose model standards.

2. Title insurance companies agree that all future policies be accompanied by a survey plat or plan; and the American Land Title Association and the American Bar Association propose model standards.

3. All title insurance surveys be recorded for the benefit of abutters and future users; and the American Bar Association and the American Land Title Association propose model standards.

4. *All boundary-survey plans show deed references of land owners and adjacent land owners until a parcel-identifier system has been adopted.*

4.1.2 Sequence of Tasks

The first task in preparing the cadastral overlay will be to establish the quality of the existing information and, where information discontinuities exist, to carry out supplemental surveys. This task should be placed under the direction of an experienced land surveyor, well versed in the peculiar nuances of the law and practice of surveying in the region. The compilation of this information will consist of the creation of a hierarchical graphical framework, holding the more highly weighted information fixed and fitting the lower-weighted information to it. Although the weighting of evidence will depend in part on the law and practice of surveying as specifically related to the region, in general it will follow the broad categories as listed in descending order below (McLaughlin, 1975):

(a) Natural boundaries as plotted on the base maps (such as lakes, rivers, and roads) will generally form the highest-weighted information framework;

(b) Geodetically referenced cadastral surveys that can be plotted on the base maps will provide the next highest weighted information;

(c) Monument referenced cadastral surveys will then be fitted to the framework;

(d) Physical evidence of original surveys (such as old rural fences) will next be fitted to the framework;

(e) Deed descriptions (which in many areas will form the bulk of the existing information) will then be fitted to either natural boundaries or to survey measurements; and

(f) A final category of information to be fitted will include deeds that merely describe abuttals, assessors' descriptions, and other similar information.

The ranking of deed descriptions below physical evidence of older surveys is based on the widely accepted legal surveying principle that the boundary as acknowledged and adhered to by contiguous landholders should have legal precedence over mathematical descriptions.

The above broad listing of categories for weighting information must necessarily be tempered by considerations of the time during which the information was first collected, the techniques and instruments used to collect the information, and corrections that have or have not been made to the information, such as the correction from magnetic to grid orientation. It is because of these considerations that the services of a professional land surveyor will be required during the compilation tasks.

Once an interim parcel base has been compiled, parcel identifiers will be assigned to each parcel. The assignment and control of these identifiers is described below. Subsequently, both the parcel locations and the assigned identifiers must be checked to ensure that

(a) All cadastral parcels within the jurisdiction have been accounted for;

(b) The best available information has been employed to determine the approximate size, shape, and location of each parcel;

(c) The correct parcel identifier has been assigned to each parcel; and

(d) Indexes are available that facilitate future access to all the documents used to locate and describe property corners and monuments shown on the maps, i.e., that the documentation is keyed to parcel identifiers, point identifiers, or some other references that appear on the maps.

While all aspects of the checking operation will be important, particular attention must be focused on ensuring that all parcels are accounted for. Only in this manner can the subsequent iterative operations be carried out reliably. Usually, the best available tool for accomplishing this task in most jurisdictions will be the current assessment record. However, experience has shown that as many as 20 percent of the parcels compiled in a cadastral overlay program may never have been shown on the assessment record. If any uncertainties are uncovered during the checking operation, the draft parcel map should be routed back to the compilers. In some instances, additional field surveys may be required.

On completion of the checking process, the overlay can then be published as part of the cadastral-mapping series. At this stage, the interim sheet should carry the date of compilation and should note that "The compiled parcel information is of a provisional nature only and must not be used for legal purposes." This declaration is necessary to ensure that this interim mapping product, which is based on information of mixed quality, is not misused. Once a cadastral survey program is implemented, the means of providing for the continuous updating and improvement of the overlay series will be available and the provisional declaration may be dropped.

4.2   CADASTRAL SURVEY REQUIREMENTS

The effective maintenance of a series of cadastral maps and, indeed, the successful implementation of a multipurpose cadastre, will be dependent in large measure on the quality of the cadastral survey system.

4.2.1   Scope of Standards Required

A cadastral survey system governs the creation and mutation of parcel boundaries. The system also maintains both microlevel and macrolevel spatial records of current land-tenure arrangements. The functions of a cadastral survey system will be defined in one or more surveying and boundaries statutes and in regulations pursuant to these statutes. In a comprehensive system, statutory authority will specify

(a) The geometric reference framework to which all information must be referred,

(b) The type and weighting of information that must be provided in evidence of the creation or mutation of a boundary,

(c) The standards of survey practice that must be met in providing this information,

(d) The authority vested in a public survey administrator to examine and register proposed boundary mutations, and

(e) The right of judicial appeal from administrative decisions.

In any jurisdiction introducing the multipurpose cadastre concept, boundaries should be recorded within the cadastral survey system to be legally effective. Furthermore, any statute that relates to the creation or mutation of parcel boundaries, such as subdivision, expropriation, quieting of titles, highway alignment, and similar legislation, should make reference to the examination and registration requirements of the cadastral survey system.

Standards for cadastral surveys may be formulated with respect to identifiers for all boundary points, monumentation (materials, dimension, reference points), information required on monuments (surveyor's name, monument number, dates), spatial accuracy of location data, data required in the record of each boundary segment (identities of end points and identities of parcels bounded), plans or plats of survey (seals, detail, cartography, approvals, materials), field books, and oaths. In this section, however, we will restrict our discussion to spatial accuracy considerations.

The legal surveying component of the cadastral survey system generally entails the two-phase operation of (a) gathering, interpreting, and weighting pertinent information and (b) spatially referencing the information. Accuracy specifications are necessary for both of these phases. There is always the danger that while analyzing the accuracy of spatial referencing the importance of accurately interpreting the boundary evidence itself will be overlooked. For example, spatial referencing may appear to relocate a boundary within a tolerance of 0.01 ft (based on an analysis of the measurements), where, in fact, because of a misinterpretation of the evidence, the boundary may be displaced several feet.

4.2.2   Accuracy of Position

Boundaries may be described by points or corners, straight lines, and/or curvilinear lines. Accuracy specifications may be expressed in terms of traverse misclosures or boundary tolerances. Of these the boundary tolerance is a far superior, if somewhat more complex, approach that is illustrated in the following cadastral relocation example (McLaughlin, 1977). In this example two cases can be described (see Figure 4.1):

1. Point P was originally coordinated and marked in the location P_1 by Survey I connected to control points A and B. Later, Survey II, connected to control points C and D, was requested in order to relocate point P in its original location corresponding to the coordinates of point P_1. Owing to the accumulation of random errors, Survey II determined that point P should be located at P_2. The question then becomes What maximum distance (maximum relocation error) between P_1 and P_2 may be expected at a given confidence level?

2. The same point P, marked on the ground, was independently coordinated by Survey I and Survey II. Owing to the accumulation of random errors, two sets of coordinates have been obtained for the same point, raising the question, What maximum differences ΔX and ΔY may be expected at a given confidence level?

In both cases there is an accumulation of errors from three sources:

   (a) Influence e_c of relative positional errors of the control points,
   (b) Influence e_o of errors of the original survey (Survey I), and
   (c) Influence e_r of errors of the relocation survey (Survey II).

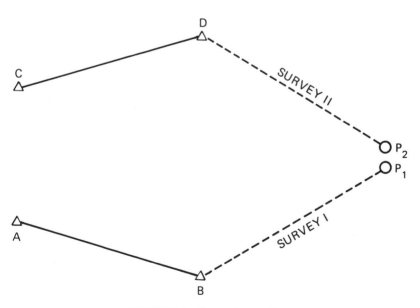

FIGURE 4.1   Cadastral relocation.

The total positioning error of P, which in the first case is expressed by the maximum expected distance $P_1 - P_2$ and in the second case as the maximum expected differences $\Delta X = X_2 - X_1$ and $\Delta Y = Y_2 - Y_2$, is a function of the three random error sources e_c, e_o, and e_r. If Survey I and Survey II are tied to the same control points, then the influence of e_c is 0. In Case 1, the maximum expected distance $P_1 - P_2$ may be expressed as the semimajor axis of a relative error ellipse between P_1 and P_2. Since the distance is treated as univariate, the semimajor axis of a standard error ellipse will correspond to a 68 percent (1σ) probability that the distance will not be larger than the value of the semimajor axis. To obtain a higher probability level (e.g., 95 percent) the semimajor axis is lengthened (statistically speaking). In Case 2, the expected maximum differences ΔX and ΔY can also be expressed in a statistical fashion at a desired confidence level. In both cases one has to determine the variance-covariance matrix $\Sigma_{\bar{x}_p}$ for point P, which is treated as two separate points P_1 and P_2. The matrix $\Sigma_{\bar{x}_p}$ is included in the general variance-covariance matrix $\Sigma_{\bar{x}}$, which contains the quality information for all coordinates involved in the survey.

To be able to estimate the achievable accuracy in relocation surveys the following information must be available for each individual case:

(a) Variance-covariance matrix of the control points,

(b) Type and standard deviations of observations in the original relocating surveys, and

(c) Configuration of the survey network that connects point P with the control points.

Little work has been done to date on formulating accuracy specifications using the approach described above. A committee of lawyers and surveyors addressing the problems of cadastral standards noted that, while the function of a modern cadastral system should be to respond to the needs of the community, the actual control over the type and quality of the product prepared has largely been a function of instrumentation capabilities and historically accepted professional practices (North American Institute for Modernization of Land Data Systems, 1975). The committee concluded that

Cadastral survey standards have for the most part been based upon the capabilities of existing technology, and have been modified in response to technological change. Furthermore, the type of standard emphasized has invariably been that which conformed with existing practices. The reliance of relative precision criteria, for example, probably reflects more on the preferences of the land surveyor than on any articulated consumer requirement. The promotion of technology with ever increasing standards may not be warranted unless acceptable marginal utility can be shown. At present we do not have a sufficient understanding of consumer requirements and preferences, and we do not know how these preferences change over a period of time as the use and importance of the land changes. (Chatterton and McLaughlin, 1975)

In a study carried out for the Maritime Provinces of Canada in 1977, it was argued that cadastral accuracy standards should be in the order of ±0.1 ft maximum error in urban areas, ±0.3 ft in suburban areas, and ±1 to ±2 ft in rural areas (McLaughlin, 1977). Much more research is required, however, before definitive accuracy standards can be put forth.

4.3 STANDARDS FOR ASSIGNING PARCEL IDENTIFIERS

The parcel identifier may be defined as a code for recognizing, selecting, identifying, and arranging information to facilitate organized storage and retrieval of parcel records. It may be used for spatial referencing of information and as shorthand for referring to a particular parcel in lieu of a full legal description.

Three important forms of parcel identifiers may be distinguished: name-related identifiers; abstract, alphanumeric identifiers; and location identifiers. Cadastral information may be retrieved through one or more of these indices. In a name code, parcel records are associated with individuals and legal entities claiming an interest to a parcel of land. An important example in current use is the alphabetical grantor-grantee index. The alphanumeric code, on the other hand, is often a random number associated with the parcel. Perhaps the simplest example is a tract index based on a sequential numbering system. Finally, a location identifier may also serve as a record index. Examples include identifiers related to the Public Land Survey System and to geographical coordinates.

Location identifiers in turn may be subdivided into at least three broad categories: hierarchial identifiers, coordinate identifiers, and hybrid identifiers. The hierarchial identifier is based on a graded series of political units, as, for example, federal, state, county, town, and ward. Within the smallest territorial unit, random parcel codes may be assigned. The coordinate identifier relates a parcel to a reference ellipsoid either through the use of geodetically derived latitude and longitude or through the use of plane coordinates. Any point within the parcel, or any conventionally defined boundary point, may be chosen for the assignment of the coordinate identifier. The location index may also employ some combination of hierarchical and coordinate coding to form a hybrid index.

4.3.1 Criteria for Designing a System of Identifiers

Parcel identifiers must be considered in the design of the multipurpose cadastre both from the standpoint of initial selection and subsequent control. There is general consensus that the record code should exhibit at least the following attributes: uniqueness, simplicity, flexibility, permanence, economy, and accessibility. The requirement of a unique index for each parcel of land is the most basic criterion and is a necessary prerequisite to the development of any multipurpose cadastre. Simplicity suggests that the identifier should be easily understandable and usable to the general public (or at least to that segment of the public that may have cause to use the system).

The identifier should be reasonably flexible, capable of serving a variety of different uses. The criterion of reasonable permanence suggests that the indexing system should not generally be subject to change and disruption. Economy relates both to the initial implementation costs and to the ongoing operational costs of the cadastral system. Accessibility refers to the ease with which the index code itself can be obtained. If a coordinate or hybrid location identifier is used as the record index, special attention must be given to the quality of the reference datum and to the accuracy of point determination.

Given the advantages and disadvantages of each indexing system, the specific choice of an optimal approach has proven difficult. The goal of the Atlanta Conference on Compatable Land Identifiers—The Problems, Prospects, and Payoffs (CLIPPP), for example, was to "recommend a single, compatible land parcel and point identifier system that would facilitate the collection, storage, manipulation, and retrieval of all land-related data" (Moyer and Fisher, 1973). This conference favored as the parcel and point identifier for universal use a number consisting of the state code number, the county code number, and the State Plane Coordinates for the point or the visual center of the parcel (Cook, 1982). Such an approach is employed, for example, in the North Carolina Land Records Management Program (North Carolina Dept. of Administration, 1981).

While the advantages of a coordinate-based identifier are well documented in the CLIPPP proceedings (Moyer and Fisher, 1973), experience suggests that the choice of a parcel index for the multipurpose cadastre in its initial stages will be dictated by local needs and resources (particularly the need for maximum accessibility and for effective administration). Nevertheless, recent developments in the software of data-base-management systems and the increasing use of multiple indices through cross-index tables permit the use of a family of parcel identifiers. The most important criteria for the primary parcel identifier at this stage are uniqueness, simplicity, and economy of maintenance. However, in the broader view, for the efficient aggregation of information related to parcels that exist in one or more cadastral systems, a uniform parcel identifier, or a set of compatible identifiers, will have to be adopted for all parcel-related data files.

What is especially important is that one of the parcel identifiers be institutionally recognized and legally defined. To ensure and facilitate its use, one parcel identifier in the land-parcel register should become the official, legal reference to all title documents affecting that parcel, that is, the general index number used by the recorder of deeds, at least for all documents filed on or after the date the register becomes available. Use of this parcel identifier would be sufficient for legal descriptions of parcels, as proposed in the Uniform Simplification of Land Transfers Act (National Conference of Commissioners on Uniform State Laws, 1977).

In addition to the legally defined cadastral parcel identifier, provision should be made to accommodate secondary indexes based, for example, on street addresses, names, or geographic codes through the use of cross-referenced indexes.

4.3.2   Control of New Parcel Numbers

While the choice of a parcel-identifier system is of fundamental concern, attention must be directed to the allocation and subsequent control of these identifiers. Any parcel-identifier system can only work if one agency has the sole authority for assigning identifiers. This preferably should be that agency responsible for land registration.

During the course of implementing the multipurpose cadastre, the parcel identifier should be introduced during the first phase of the cadastral-mapping program. Assignment of a parcel identifier should be provisional until the cadastral maps are formally approved and registered.

The control of the subsequent allocation, re-allocation, and withdrawal of parcel identifiers is but part of the larger process of managing land-tenure changes. If the configuration of a parcel is changed (e.g., typically by subdivision), that parcel ceases to exist, and new identifiers are assigned to the new parcels. However, the original parcel remains as a historic entity, and the descriptions of it that were entered in the various registers and files when it did exist remain coded to it. Indexes that identify such "retired" parcels must be included in the records system unless provided otherwise by statute.

A special problem with the parcel-identifier system concerns the possibility of error in data processing. There are various techniques available for monitoring the fidelity of a parcel identifier from the time of its initial assignment through subsequent processing, most of which employ the addition of a redundant check digit. The simplest approach is the addition of a check digit at the end of an identifier, which is mathematically related to the sequence of digits in the identifier.

5
Organizing Other
Land-Parcel Records

Cadastral record systems or files, as previously defined, are parcel oriented. Each cadastral record contains, in addition to other information about the parcel in question, a parcel identifier that is unique to that parcel. The identifier provides an ideal link to interrelate the many other files that contain information about the same parcel. The benefits of having a single agency responsible as the source for each element of land data in the public records are manifold. A cadastre that serves as this multipurpose link in a land-data system must meet a variety of requirements of other users beyond those described in Chapters 2, 3, and 4.

Currently, the most extensively developed cadastral record systems in North America are the essentially single-purpose systems of local land-title recording offices and of assessors. Widespread initiatives to modernize these systems, usually by computerizing them, have led to numerous examples of their being used for more than one purpose. The adaptation of these single-purpose cadastral record systems to other purposes also has spurred interest in systems that were designed from the outset to serve multiple purposes. Conditions seem favorable for such initiatives to succeed: adequate computing power is within the technical and financial grasp of the smallest and poorest local government, the pool of talent to design and install systems is growing, officials in other governmental functions—notably planning and public works—are becoming increasingly aware of the potentialities, and the pressures to increase the cost-effectiveness of local government are growing.

The objective of this chapter, therefore, is to provide information on standards and practices that will facilitate (1) the joint development, multiple use, or both of cadastral records by local governments and other local users of land information and (2) the establishment of other registers of land-parcel data compatible with the cadastre that can serve the broader needs of regional, state, and national government agencies, public utilities, and others.

Land-record systems can require the gathering and processing of vast amounts of information from numerous sources. This information is used to locate and identify parcels, describe them and the buildings erected upon them, and meet the specific needs of the users of the records.

5.1 MEASURING USER REQUIREMENTS

A multipurpose cadastre, as its name implies, is designed to serve more than one purpose. This means that the informational requirements of more than one set of users need to be taken into consideration in the design of the cadastre. It follows that these needs should be borne in mind in the design of any cadastral record system, even though it may be designed to serve only a single purpose.

5.1.1 Nature of Interests in Cadastral Information

While attention to user requirements is an obvious aspect of system design (see Section 5.4), the analysis of user requirements may present some special problems in the design of a multipurpose cadastre. The essential task is to determine the degree of commonality of interest in a particular data element on the part of the potential users of the information in a cadastre. Those data elements for which there is widespread interest might be maintained in an integrated cadastral record system, while those of limited interest might be maintained in a specialized system.

The study of the commonality of interest in data elements is complicated by the fact that potential users of the information in the cadastre may not be readily identifiable. Attempts to identify potential users must take into account differences in the organizational structure of local government (i.e., the existence of different types of general governments—such as counties, municipalities, and townships—and special governments—such as school districts and fire-protection districts); the functions assigned to each unit of government; and the organization and nomenclature of offices, departments, and other organizational units. Listed below are some local government offices (variously named) that are potential users of cadastral information:

Administration	Finance (treasurer, tax
Assessment	collector)
Building inspection (code administration)	Fire and emergency medical
Clerk	services
Court administration	Forestry
Deed recordation	Health (disease control)
Engineering	Housing code enforcement

Natural resources

Parks and recreation

Planning

Police

Pollution control (air, water,
hazardous wastes)

School planning and districting

Streets and highways (traffic control)

Surveying

Transportation

Utilities (electric, gas, street
lighting)

Voter registration

Water and sewers

Zoning

Other lists are found in *Guidelines for Systems Analysis of User Requirements* (American Public Works Association Research Foundation, 1981a) and in *Monitoring Foreign Ownership of U.S. Real Estate: A Report to Congress* (U.S. Department of Agriculture, 1979).

The actual information needs of these users may be difficult to ascertain. Many users, such as planners, are accustomed to having to make use of suboptimal land information. In other instances, a single characteristic customarily may be viewed in different ways. It appears that these issues have not been rigorously investigated.

Listed below are some of the areas in which user requirements may differ. Designers of cadastral record systems should explore each of these areas with each potential user of a multipurpose cadastre.

Spatial Interests. Although all users of a multipurpose cadastre may be assumed to have an interest in parcel-oriented information, they may have other, sometimes more important interests. For example, they also may be interested in areas within parcels, areas larger than parcels, or areas that overlap parts of parcels. Frequently, the interest is with buildings, which may be smaller than, larger than, or overlap the boundaries of a parcel. Larger areas of interest may be political, economic, or ecological in nature. Sometimes the interest is in relating parcels to arbitrary grid cells.

Temporal Interests. The configuration of parcels, their owners, and their characteristics ordinarily changes with the passage of time. Some users may be interested in the current situation. Others may be interested in the situation at specified intervals. Still others may be interested in the history of the changes or in summaries of changes over specified intervals such as calendar years or fiscal years.

Coverage. Requirements with respect to the coverage of parcels also may vary. Many users require information on all parcels in an area. Some users, however, may only be interested in certain parcels. Sometimes those parcels are determined by the course of events. Other times, the parcels of interest may be selected by a sampling procedure, although such an interest implies a need for some minimal amount of information on all parcels.

Subjects. The subjects (things of interest about parcels) will vary greatly among users. The interest may be with the physical characteristics of the parcels themselves, the physical characteristics of buildings and other improvements to the parcels, uses of parcels, events (such as fires, crimes, or illnesses), transactions (such as sales or leases), the parties to such transactions, or some combination of subjects.

Precision. Users may have differing requirements concerning measurements and descriptors. Generally, obtaining greater precision requires greater expenditures for data collection and maintenance.

Linkages. Although the linchpin of a multipurpose cadastre composed of a number of separate cadastral record systems is the parcel identifier, some users are primarily interested in other, nonunique identifiers, such as street address, name (of a person or an establishment), or geographic location (relative to either a specific geographic feature or a reference framework).

Accessibility. Just as informational requirements vary, so do accessibility requirements. Some users may require immediate access to a record, document, or image. Others may be satisfied with access within a few minutes, while still others may be satisfied with overnight or even weekly access service. Among the factors that affect accessibility requirements are the volume of information that is being processed and the value of the time of the users.

5.1.2 Requirements for Land-Title Recording

Maintaining land-title records is a function of county government in the United States except in three New England states (Connecticut, Rhode Island, and Vermont), in which it is a function of city and town governments (Almy, 1979a). Hence, there are about 3000 county land-title record systems and about 500 city and town systems.

Typical and innovative land-title recording and registration practices are described in some detail in *American Land Title Recordation Practices: State of the Art and Prospects for Improvement* (U.S. Department of Housing and Urban Development, 1980). That report describes innovations in 13 localities, and a subsequent report, *Profiles of the Land Title Demonstration Projects* (U.S. Department of Housing and Urban Development, 1981a, 1981b), describes HUD-funded efforts to improve land-title recordation and registration procedures in nine localities.

A typical land-title recording system is a register of evidence of title to real property, such evidence being contained in copies of deeds, land contracts, wills, and other documents. Title records, therefore, essentially are parcel oriented, and all legally recognized parcels are included in the system. A land-title record system is an archive that makes it possible to track changes in the configuration of parcels and to construct a "chain of title." Such activities often are made difficult in conventional, manual systems because of the cumbersome way in which documents are indexed, and access to documents can be streamlined in a number of ways that are discussed in Section 5.3 and Chapter 6.

In a conventional, manual system, access to individual documents usually is obtained by (1) searching alphabetical indexes of the names of buyers (grantees) and sellers (grantors) or by searching a tract index, which essentially is a subdivision index; (2) noting the location of the documents of interest in the register; and (3) reading them or obtaining copies. The location of a particular document in a register often is indicated by a numerical identifier that refers to the appropriate volume and page of the register. Other times, the reference is to a serial document number. Organizing documents in the sequence in which they are received is administratively efficient. Users of land-title records, however, often only possess information on the name of a person, the address of a property, its legal description, or an assessor's parcel identifier. Thus, use of land-title records is facilitated if there are name, address, parcel-identifier, and legal description indexes available to locate the documents. Such indexes make it possible to link land-title records to other cadastral records.

A number of technologies can facilitate access to land-title records. Micrographics and video technologies can be used to store copies of title documents, and photocopy equipment can be used to produce copies of documents on demand. Indexes can be computerized. However, it may not be economically feasible to computerize current land-title records in their entirety, let alone historic documents, since record conversion would be a monumental task. Computer storage of title documents does become feasible when the documents are standardized.

Recording officers may be responsible for maintaining property-ownership maps, assigning parcel identifiers, and transmitting information on changes in ownership and sales prices to assessors and others.

5.1.3   Requirements for Real-Property Assessment

In the United States, local governments are primarily responsible for real-property assessment except in Maryland and Montana, where the states are fully responsible for assessment (Almy, 1979a). There are approximately 13,400 county, municipality, and township assessment jurisdictions.

Although specific responsibilities are set forth in state law, assessors generally are responsible for (1) locating and describing properties, (2) estimating their values, (3) linking them to their current owners, and (4) designating their official value for tax purposes, taking into account legal reasons for assessing them in amounts that differ from their appraised values. Accordingly, assessors must collect, store, retrieve, and analyze a great deal of information that is parcel oriented (see Table 5.1), and all parcels should be included in a real-property assessment system. Data requirements are described in more detail in *Improving Real Property Assessment: A Reference Manual* (International Association of Assessing Officers, 1978), which is the source of much of the material in this chapter, and a list of recommended data elements for a fiscal cadastre (assessment and taxation) can be found in *Multiple-Purpose Land Data Systems, Monitoring Foreign Ownership of U.S. Real Estate* (Moyer, 1979). While it is administratively convenient to revise property records as changes occur, assessors officially are concerned with the status and value of parcels on the legally designated annual appraisal or assessment date. Thus, assessors have only a limited interest in information that is of a historical nature.

Table 5.1 Contents of Assessment Record Systems

Data Element	File			
	Legal Description (Maps)	Property Characteristics (Property Records)	Market Data (Sales, etc.)	Property Ownership (Assessment Roll)
Boundaries of individual parcels	X			
Parcel dimensions and/or areas	X	X	X	
Bearings (where applicable)	X			
Subdivision names, boundaries, lot numbers, etc.	X	X	X	
Governmental boundaries	X			
Easement and right-of-way boundaries	X			
Location and name of streets, etc.	X			
Assessors' parcel identifiers	X	X	X	X
Street address	X	X	X	X
Property-use classification code		X	X	X
Assessment-status code		X	X	X
Tax-rate-area code		X	X	X
Neighborhood or market-area code		X	X	
Site characteristics		X	X	
Improvement characteristics		X	X	
Building perimeter sketch		X		
Building-cost data		X		
Income and expense data		X	X	
Building permit history		X		
Sale history		X		
Sale date		X	X	
Sale price (nominal)		X	X	
Cash-equivalent sale price			X	
Time-adjusted sale price			X	
Sale acceptance/rejection code			X	
Source of sale confirmation			X	
Type of instrument			X	
Instrument number			X	
Assessment/sale price ratio			X	
Appraised and assessed values		X	X	X
Record of on-site inspections		X		
Appeals history		X		
Appraiser's name (coded)		X		
Year appraised		X	X	
Name of owner				X
Address of owner				X

Depending on the nature of the information and the degree to which assessment records have been computerized, those records may be maintained in a number of separate files.

Legal Description Files (Maps). The assessor's primary working "file" containing legal descriptions is a set of property-ownership maps, in which legal descriptions are represented graphically, not merely in writing (see Chapter 4). Only by representing the size and shape of parcels graphically can the assessor be sure that all taxable property is assessed and that none is assessed twice. Knowledge of the size, shape, and location of parcels is essential to land appraisal. Maps also are indispensable in other aspects of appraisal operations and can serve many other purposes as well. For these reasons, it is often stated that the first requirement of a good assessment system is a complete set of up-to-date property-ownership maps.

Legal descriptions are generally printed in abbreviated form on assessment rolls, and they often are found on individual property records. Maintenance of complete legal descriptions in an assessor's computerized records usually is regarded as a wasteful use of computer storage.

Property-Record Files. The largest file in a good real-property assessment system is the property-record file. This file contains information describing each parcel and any buildings or other improvements on each parcel. It also documents the factors and methods used in appraising each property. The specific items of information contained in a property-record file are determined by the information required to identify and describe each property, to make appraisal calculations, and to satisfy the owner(s) of the property that the assessor is familiar with it. Property-record contents also are heavily influenced by the type of property in question, the specific appraisal method or methods that the assessor uses, whether the file is computerized, and regional factors such as climate and culture. Some typical site and improvement characteristics, not listed in Table 5.1, are listed in Table 5.2 (other characteristics are identified in International Association of Assessing Officers, 1978, and Moyer, 1979). The file essentially is a record of the current status of properties, although it is not uncommon for assessors' property records to contain a 5- to 10-year history of assessments, sales, building permits, and appeals.

Assessors' property records are examined and revised virtually on a daily basis. Property-record information is used or processed whenever a property is scheduled for a visual inspection, a property is reappraised, a property is sold, a building permit is issued for the property, a property is damaged by fire or other disaster, a property owner or other person wishes to examine the record, or a property's assessment is appealed or is reviewed by a review agency. The volume and frequency of inquiries and revisions have led many assessors to place their property records in on-line computer files. The more typical paper files of property records usually are arranged in parcel-identifier order, but they may be arranged by address, subdivision, or owner's name.

TABLE 5.2   Site and Improvement Characteristics

Site/Location	*Construction Quality*
Topography	Quality of materials
Soil characteristics	Workmanship
Usable land area	Architecture
Building setback requirements	
Landscaping	*Design*
Cul-de-sac location	Intended use
Corner location	Architectural style
View	Shape of building
Street and alley access	Roof type
Railroad and waterway access	Story height
Available utilities	
Distance to shopping, etc.	*Other Building Features*
Nearby nuisances	Number of rooms by type
Zoning	Heating, ventilation, air conditioning
	Plumbing facilities
Building Size	Fireplaces and similar amenities
Ground-floor area	Additions and remodeling
Total floor area	Porches and patios
Leasable area	Swimming pools
Volume	Shelter for automobiles
Building height	Elevators
Ceiling height	Power equipment
Clear span	
Number of stories	*Age/Extent of Depreciation*
Number of units (apartments, etc.)	Chronological age
	"Effective" age
Shape	Remaining economic life
Floor area/perimeter ratio	"Percent good"
Number of corners	Condition
	Extent of remodeling
Construction Materials	
Foundation	
Framing	
Floors	
Walls (exterior and interior)	
Ceilings	
Roofs	

Market-Data Files. Assessors often maintain specialized market-data files. These files contain data on sales prices and terms, rental revenues and operating expenses, and building costs and data on property characteristics as of the date of sale, the dates to which the rental and operating expense data applied, or the date construction was completed. While property-record files may contain identical kinds of information, the distinction between the two types of files is important. Property-record files contain status information on all properties, while the special market-data files contain descriptions of only those properties for which the market data are available and of characteristics and conditions in existence at the time of sale, rental, or construction. Sales data and other market-data files are crucial to the development of the valuation models used to appraise all properties. Sales files may be computerized, or they may be organized similar to manual property records. Income and expense files and cost files are usually more informal since fewer properties are involved.

Property-Ownership Files. The assessment roll normally is the assessor's primary ownership file. Nowadays, assessment rolls usually are computerized, and assessment-roll information may be available only on computer screens or on microfilm. Assessment-roll entries normally are arranged in parcel-identifier order and contain the name and address of the current owner or taxpayer; the legal description of the parcel (often in abbreviated form); the assessed value (often separate land and improvement values are listed); and such additional information as a property-use code, a tax-rate area code, tax extensions, and the amount of exemptions applying to the property. Owner's or taxpayer's name, address, and subdivision indexes are often prepared. Exemption applications are contained in ancillary files.

The investment required to collect assessment data (particularly property characteristics data) and to convert them into usable form is substantial. If the data are maintained, if the data base is sufficient to accommodate new and changing needs, and if the data are accessible, this investment can be shared by a number of offices and amortized over a long period of time.

5.1.4   Requirements for Land-Use Planning and Regulation

Land-use planning and regulation are essentially local government functions. Because of the diversity of planning issues and land-use control techniques, the functions are not easily defined, but they generally deal with identifying and applying ways of guiding development and use of the physical environment that promote the health, safety, welfare, and convenience of the citizenry. Planning, therefore, deals with monitoring changes in land use and in the physical environment, identifying problems, proposing solutions, and attempting to implement the most feasible solution to the problem in question.

Land-use planners require population, economic, social, and environmental data. Planners are increasingly likely to have data systems that are designed to meet their particular needs, examples of which are described in *Computers in Local Government: Urban and Regional Planning* (Auerbach Publishers, 1980a) and in the October 1981 issue of *Planning*, which is published by the American Planning Association, Chicago, Illinois.

Cadastral records can be used in a number of planning activities, such as making land-use inventories; monitoring development; evaluating proposed developments;

Table 5.3 Examples of Cadastral Data Elements Used in Planning

Site/Location Characteristics	*Land-Use Data*
Street address	Land-use code(s)
Geographic location	Business license history
Political subdivision codes	Number of residents
Parcel dimensions and/or areas	Number of employees
Easement data	On-site parking spaces
Street and alley access	Zoning
Railroad and waterway access	
Available utilities	*Improvement Characteristics*
Distance to shopping, etc.	Story height
Topography	Floor-area ratio
Soil and subsoil characteristics	Dwelling units
Groundwater and surface-water	Other units
characteristics	Condition of buildings
Natural vegatation	Building permit history
Presence of air and water pollution-	Landscaping
causing agents	

selecting sites for public and quasi-public facilities (such as schools, fire stations, hospitals, and power plants); making transportation plans; and delineating zoning, redevelopment, rehabilitation, historic, and other districts. The administration of zoning and subdivision ordinances is parcel-specific. Urban-design activities often are parcel oriented. Cadastral data elements that are used in planning include those listed in Table 5.3.

5.1.5 Requirements for Public Works

Public works administration involves activities that relate to the design, construction, maintenance, and operation of public buildings, utilities, transportation systems (including streets and roads), and other facilities. Public works officials may be responsible for the analysis of the need for such capital improvements, and they may be responsible for selecting and acquiring sites. In these respects, their needs of cadastral data are the same as planners. Once sites have been acquired and the infrastructure has been constructed, there is a need for information on the physical characteristics and precise location of the infrastructure relative to rights of way, other utility systems, and abutting parcels, as well as for information on changing service demands. A particular concern has been mapping and maintaining records on underground water, sewage, electricity, gas, telephone, and other systems. The best source of information on these subjects is the *Computer Assisted Mapping and Records Activity System Manual* published by the American Public Works Association, Chicago, Illinois (1980-1981).

5.1.6 Requirements for Public Health and Safety Functions

Several governmental functions involved in protecting the health and safety of the populace originate or use parcel-oriented data. Building-code administrators are responsible for ensuring that buildings meet safety standards. In fulfilling this responsibility, building-code administrators review demolition, excavation, construction, and alteration plans and issue permits if the plans conform to the code. In the process, they collect copies of architectural drawings and specifications that contain information on the size of structures, construction materials, and construction methods. This information can be useful to assessing officers, police officers, and fire fighters. Assessors also use building-permit information to alert them to possible changes in property characteristics. Information on the number and value of building permits is used in some economic studies. Building inspectors must maintain systems for keeping records of the status of each permit that has been issued as well as of certificates of occupancy, which are issued when new construction or remodeling meets safety standards, and of records of violations of the code. Information on violations is useful in a variety of other governmental functions. Permits and similar documents are usually filed in permit number order, although cross-references to legal descriptions and addresses are useful in locating buildings in the field.

Cadastral records also are used by police, fire, and health departments. Such records are used in identifying potential hazards, in selecting sites for fire and police stations, and in the monitoring of septic and sewer systems.

5.1.7 Requirements for Financial Management

Cadastral records are used in at least two aspects of financial management: real-property taxation and fiscal-impact analyses. The real-property taxation function takes up where the assessment function leaves off (and some finance departments are responsible for the assessment function, while some assessors are responsible for property-tax collection). The taxation function includes (1) the issuance of property-tax bills, (2) receiving tax payments and maintaining records for those payments and amounts due, (3) maintaining a record of tax liens, and (4) instituting enforcement procedures when taxes become delinquent. Real-property taxation systems may be integrated in one way or another with real-property assessment systems and with governmental financial-management systems.

Fiscal-impact studies often are both a financial-management concern and a planning concern. Hence, they may be performed by either finance or planning departments. When the studies involve specific development proposals or changes in the property-tax base, property-tax rates, or property-tax policies, cadastral data are required in the analyses. Perhaps, the best reference on fiscal impact studies is *The Fiscal Impact Handbook: Estimating Local Costs and Revenues in Land Development* (Burchell and Listokin, 1978).

5.2 STANDARDIZING THE DESCRIPTIONS AND CODING OF PROPERTY CHARACTERISTICS

The sharing of land-parcel data among the users of a multipurpose cadastre will depend on their use of common procedures for describing and coding property characteristics. Describing a property characteristic involves a depiction in words or a representation by a picture. One may draw sketches, take photographs, take measurements, make counts, compile lists, sort into classes, or assign ratings. The choice as to which techniques to employ depends on such factors as the nature of the characteristics being described and the purposes for which the data are being collected. Coding is the reduction or abbreviation of a description to a more manageable size through the use of letters, numbers, symbols, and fewer words. For universal use of the data, these codes must be standardized.

5.2.1 Alternatives for Classifying Land Parcels

Descriptions of land-parcel characteristics may be objective or subjective. Subjective descriptions require more intellectual effort, whereas objective descriptions are made more mechanically. Similarly, a description may be qualitative or quantitative. A house may be described as a mansion because mansions are "large" and the house in question is the largest house around—a qualitative description. Or the house may have been classified as a mansion because an assessor's cost manual specifications indicate that, among other things, mansions must have a ground-floor area equal to or greater than 3000 square feet, and the house in question has a ground-floor area of 3130 square feet—a quantitative description. There is often a close correspondence between subjective and qualitative descriptions and between objective and quantitative descriptions.

Property characteristics may be continuous, discrete, or dichotomous. A continuous characteristic or variable is one that may take on any numerical value. Building area, for example, is a continuous variable. A discrete variable, such as number of fireplaces, can take on any whole number value, usually within certain limits. The number of rooms in a dwelling ordinarily would be thought of as a discrete variable, although "half-room" counts are sometimes used. Dichotomous characteristics or variables are those having to do with the presence or absence of a condition. They are usually described by answering a yes or no question. The variables arising from the answers to such questions are often called "dummy" variables. Examples of questions that create dummy variables are, "Does this site have lake frontage?" and "Is this property defined by the Public Land Survey System?"

Some characteristics, such as construction quality and building condition, can be treated as discrete variables by developing a rating scheme or as series of dummy variables. Building condition or state of repair, for example, could be described on a scale, of, say, 1 to 10 or as poor, fair, average, or good.

Rating schemes can be based on an absolute standard, that is, a standard that applies to all properties in the system, or they can be based on a relative standard, that is, one that changes from neighborhood to neighborhood or from one group of properties to another. If the rating standard is absolute, a building described as being in good condition in one location would also be good in any location. An example of a variable described by a relative standard is a typical or representative lot, the determination of which is based on the average size of lots in the neighborhood or area.

Coding schemes should account for all possibilities (be exhaustive), and coding categories should be mutually exclusive. In addition, property characteristics should be described and coded in a consistent way. Interestingly, when multiple regression analysis is employed in property valuation, consistency can be as important as accuracy, since the mathematical logic of the technique can compensate for inaccurate descriptions as long as they are consistently inaccurate. Obviously, consistently inaccurate data would be useless for most other purposes.

Quantitative and objective methods of describing land-parcel characteristics result in more consistent descriptions. They require explicit consideration of more details and therefore are apt to be more time-consuming, although less experienced data collectors (e.g., temporary data collectors and trainee appraisers) can be used. Consistency in the coding of qualitative or subjective characteristics, on the other hand, requires well-trained and experienced data collectors (e.g., appraisers and data-collection specialists).

5.2.2   Characteristics of the Land and Location

It would be difficult to construct a list of all characteristics of land parcels that might be indexed with reference to the multipurpose cadastre. The following are characteristics that are important to many users and for which guidelines for describing and coding may help to make the various registers of land data more compatible. Many other data elements are of interest to only one of the user departments and thus do not warrant discussion here.

Parcel Size. Parcel size may be described in terms of parcel dimensions (e.g., lot frontage and depth), land area, and usable land area. Parcel dimensions and area are obtained from surveys, plats, or maps. Usable land area is determined by reference to actual parcel dimensions and area, land-use controls (e.g., permitted coverage of the parcel—building setback and side- and rear-yard requirements), shape characteristics (e.g., extremely narrow parcels), soil characteristics, and terrain and topographic characteristics (e.g., location on a hillside or a ravine or in a floodplain). Usable area, therefore, is determined by a combination of objective measurements and subjective, personal observations. Parcel size also may be described in relative terms. For example, a lot may be described as "typical" or representative of surrounding parcels.

Land Use. Land use may be described in terms of whether the land is unimproved (i.e., without buildings) or improved and, if the land is improved, what the use(s) is (are). Unimproved land may be described in terms of actual use (e.g., agricultural use), in terms of likely use (e.g., surrounding use or permitted use), or in terms of subdivision characteristics. Coding improved land is more complex. Not only should all significant land uses be included in the scheme, but mixed uses should be accommodated. Techniques for dealing with mixed uses include (1) coding only the predominant use, (2) coding the highest and best use, (3) devising a coding system that permits secondary uses to be coded, and (4) using building-use codes as well as land-use codes. The standard procedure is to draw up a list of land uses and assign code numbers to them. See, for example, *Standard Land Use Coding Manual: A Standard System for Identifying and Coding Land Use Activities* (Bureau of Public Roads and U.S. Urban Renewal Administration, 1965). Standard property-use coding systems have been developed by assessment agencies in 33 states (Almy, 1979b), and the International Association of Assessing Officers (1981) has published a standard on property-use codes. Usually land uses are grouped according to broad classes (e.g., residential, commercial, industrial); within each major group more detailed land-use descriptions are found.

The Baltimore Land Use Coding (BLUC) System, which is a good example of a locally developed system based on the *Standard Land Use Coding Manual*, consists of a five-digit number that defines to four levels of detail the existing predominant use of a parcel of land and indicates the general secondary use. The first digit identifies the predominant use of the parcel as being in one of the following major categories: 1, residential; 2 and 3, manufacturing—two groups; 4, transportation, communication, and utilities; 5, trade; 6, services; 7, cultural, entertainment, and recreational; 8, agriculture; and 9, undeveloped land and water areas. The second digit is a refinement of the first, a subcategory of the major category. The third digit further refines the second, and the fourth digit refines the third.

As an example, code 2184 describes land used for "distilling, rectifying and blending liquors." The first digit, 2, identifies this land as being in the major category "manufacturing." The second digit, 1, identifies the land as being for the manufacture of "food and kindred products"; and third digit, 8, refines this as the manufacture of "beverage." The fourth digit, 4, identifies the beverage manufacturing as "distilling, rectifying and blending liquors." Through this four-level structure of coding, information may be retrieved to the degree of detail considered most appropriate for analysis and presentation of the data.

By addition of the fifth digit, information on parcels with a secondary use may be obtained. There are eight categories of secondary use, corresponding to the eight major categories of predominant use listed above. If, in the previous example of land being used for "distilling, rectifying and blending liquors," a wholesale trade in the product was also carried on at this location, the number 5 would be placed in the fifth-digit position. This would indicate that trade was carried on as a secondary land use.

Detroit, Michigan, has taken a different approach: in addition to a three-digit land-use code, Detroit also employs a three-digit building-use code. The land-use code always begins with zero to distinguish it from the building-use code, and the second digit is one of the following eight major categories: 1, residential; 2, commercial; 3, industrial; 4, utilities and communications; 5, transportation; 6, public and quasi-public; 7, outdoor recreation; and 8, extractive and agricultural. The third digit is a refinement of the second. As an example, code 032 describes land use for specially constructed industrial buildings.

The more complex the coding scheme, the more knowledge is required of data collectors.

Service and Transportation Network Access. Available services (e.g., access to streets or roads, alleys, railroads, waterways, telephone, electricity, gas, water, and sewers) affect the suitability of land for development and hence land value, and information about these services is useful in a number of ways. With the exception of underground utilities, these property characteristics can be observed easily and coded yes or no. Service to the property by underground utilities may be detected by the presence of street lights, meters, and above-ground connections or by recourse to utility maps. The location of utility lines eventually should be incorporated in the multipurpose cadastre.

Locational Characteristics. Locational characteristics that are external to the parcel, such as an outstanding view, the presence of a nuisance, or distance to shopping, can have important effects on land values. These characteristics are also useful for descriptive purposes. Nuisance and view characteristics usually must be measured subjectively. A rating scale that imparts a degree of consistency in describing these characteristics may be developed. Distance variables are difficult to measure but may be obtained by scaling from maps, by counting city blocks in urban areas, and by calculating distances using the geographic coordinate parcel identifiers of the parcel being described and of a parcel representing the target or reference point.

Neighborhood. It can be seen that often many of the site or location characteristics described above are common among adjacent parcels. In such situations, needless duplicate description and coding efforts are likely. One way to overcome this problem is to use the concept "neighborhood" as a generalized location variable. Neighborhood boundaries can be delineated, and factors to be considered in delineating neighborhood boundaries include land use; homogeneity of property characteristics; the presence of schools, churches, and similar cultural "magnets"; physical barriers; and trends in property values. In some areas, municipal boundaries and the boundaries of other tax-rate areas have also been found to be important neighborhood boundaries.

5.2.3 Characteristics of Structures

Building Size. Building size may be described in terms of ground-floor area, total floor area, volume, building height, number of stories, or a combination of several of these. Other, less-complete, measures of size include leasable area, ceiling height, clear span, and number of units such as apartments. In theory at least, areas, heights, spans, volumes, and units can be measured with great precision and are therefore objective characteristics. Generally, exterior dimensions are measured, since measuring interior dimensions usually is not cost-effective except in measuring ceiling heights and spans in industrial buildings. Taking building measurements is not necessarily easy, however. Vertical measurements are difficult to obtain. Curved walls and some angles are difficult to measure. Other horizontal measurements may be difficult because shrubbery, fences, and other obstacles impede the process.

Shape of Building. The shape of a building of a given floor area has an important bearing on the cost of a building, and shape characteristics are important in appraisals based on replacement cost. Shape may be described in terms of the ratio of floor area to perimeter and the number of corners or by matching the shape of the perimeter of a building with a generalized pattern (rectangular L-shaped, T-shaped, and H-shaped structures).

Design. Design characteristics can be described in terms of intended or designed use (e.g., single-family residence, gas station), arrangement of stories (e.g., two-story, one-story, split-level, trilevel), type of roof (e.g., flat, mansard), period of construction (e.g., modern, conventional, old), and architectural style (e.g., colonial, Cape Cod, ranch). Intended use can be coded in the same fashion as land use. Decisions have to be made about how to treat situations in which intended use and actual use differ (a house used as a restaurant). If the characteristic is used only to inventory land uses, current use is generally best.

The classification of story height can present problems. One is the classification of finished areas under sloping roofs. Buildings with mixed story heights are also difficult to describe.

Period of construction can be coded with considerable precision as long as the ages of properties are generally known and as long as the boundaries of periods are specified.

Architectural style is difficult to describe effectively because architectural styles are imprecisely defined, and many buildings contain elements of many styles.

Construction Quality. Like neighborhood, construction quality is a composite characteristic. It describes the cumulative effects of workmanship, the costliness of materials, and the individuality of design. With respect to construction quality ratings, an important point is that most rating schemes are designed to facilitate the use of a specific set of cost schedules used in estimating the replacement costs of buildings.

Construction quality ratings are assigned on the basis of matching a building to a set of specifications. The specifications for each class or grade should identify and describe the specific characteristics of building materials, workmanship, and other features that distinguish that class from the others. Quality-class ratings should be assigned without regard to the state of repair of the building. In other words, data collectors should assign the rating as though the building were of new construction. A knowledge of historical construction practices is helpful in assigning quality-class ratings to older buildings.

Construction Materials. The materials used in the construction of the foundations, frames, floors, walls, and roofs of buildings are required in cost estimating and in describing buildings. Many materials (e.g., wood, brick, concrete) can be observed and identified without any special training (see Section 5.3.4).

Other Building Features. Information on many other building features may be needed to describe a property adequately. The features that are important vary, of course, with property type and locality. Many of these characteristics can be described through the use of yes or no variables. Others can simply be counted.

Age/Condition. Buildings are not indestructible, and it is important to gauge the effects of aging and wear and tear on buildings. Usually both age and condition are described.

Age may be described simply in terms of chronological age or in terms of "effective age" (i.e., age adjusted for condition and remodeling). Age is sometimes described in terms of remaining economic life, which is the future period that a building is expected to contribute positively to the total value of the property. Chronological age can be determined accurately from assessment or building-inspection records but is a comparatively meaningless characteristic if condition is disregarded. Effective age and remaining economic life are, on the other hand, nebulous concepts that are difficult to estimate.

Condition, also a subjective concept, can be described in terms of a rating scale (e.g., fair, average, good) or in terms of a continuous scale (e.g., percent good).

Although standard land-use codes are available, there remains a need for a standard classification of the characteristics of structures.

We recommend that the National Association of Counties institute a project to provide the counties with a draft of a standard classification of the characteristics of structures.

Since improvements in this standard classification of the characteristics of structures will come through its use, there should be a continuing administrative unit that can update this classification and provide the necessary information to all counties.

5.3 PROCEDURES FOR COLLECTION AND MAINTENANCE OF COMPATIBLE DATA

The work of data collection and maintenance represents the capital investment in an existing register of land-parcel data. To preserve the future value of this important public asset, the collection and maintenance work must be consistent in its adherence to standard procedures, and the level of confidence in the accuracy of each sector of the data base must be known.

The importance of these procedures is underscored by the findings of a recent survey of opinions of 174 experts in computerized land-data systems. The two factors most often rated high in importance for successful implementation of a land-data system were (1) a defined responsibility for the sources and accuracy of each record and (2) standards for the quality of data that may be entered (Lincoln Institute of Land Policy, 1982).

Data-collection efforts are of two general types: initial, comprehensive data-base building efforts and data-base maintenance efforts. The former type of effort is necessary whenever new applications require additional data or whenever data-maintenance efforts have been badly neglected or are currently inadequate to keep abreast of changes in properties.

Data-base building efforts involve the following steps: (1) determining data needs (see Section 5.1), (2) developing a system for describing and coding property characteristics (see Section 5.2), (3) designing data collection forms (see Section 5.3), (4) selecting and training the data collectors, and (5) managing the data-collection efforts.

Data-base maintenance efforts are of two types: (1) efforts triggered by the issuance of building permits or similar property-specific change notices and (2) general field reviews designed to verify the correctness of current information and to detect changes that were not picked up by other means.

5.3.1 Typical Data Sources

This section identifies typical sources of cadastral data for which standard collection procedures are needed. In general, the same sources are used in both original and maintenance data-collection efforts, although differences in collection practices will be noted.

Deeds and Other Real-Property Transfer Documents. Deeds and other real-property transfer documents may provide information on (1) the identification of owners of interests in real property, (2) the nature of those interests, (3) the identification of the parcels involved (legal descriptions), (4) the type of transfer (deed, land contract, will), and (5) the prices and terms of sales or other transfers.

Building Permits. Building permits, in addition to the regulatory purposes that they serve, alert assessors and others to possible changes in the physical characteristics of buildings and other permits. The acquisition of building-permit information facilitates the timely and efficient revision of cadastral records, particularly when the contemplated building activity is described.

Planning and Zoning Documents. Planning and zoning actions (zoning changes, adoption of a master plan, urban renewal or redevelopment requirements, building-permit freezes, or sewer moratoria) may determine whether land can be developed and how property can be used, and they also influence property values. To the extent that planning actions and land-use controls directly affect individual properties, information about such actions and controls should be indexed by cadastral parcel number.

Aerial Photographs, Surveys, and Plats. Aerial photographs, cadastral surveys, and plats may provide information on the location, the size and shape, the use of parcels, and the occurrence of changes in the inventory of parcels, land uses, and improvements. Such information obviously is crucial to the origination and maintenance of cadastral records.

Information Supplied Directly by Property Owners. Property owners themselves can be highly useful sources of land-parcel data. They may be called on to verify or supplement data on property characteristics, sales prices and terms, rental income and expenses, and construction-cost data. Formerly, in assessment administration, property owners were almost exclusively relied on as the source of information on the nature, extent, and value of their properties. Such exclusive reliance resulted in the property tax being a tax on honesty, and great reliance on property owners became discredited. Recently, limitations in assessment budgets have caused assessors to revaluate property owners as a source of information. Some jurisdictions now provide property owners with detailed descriptions of their properties so that they can verify or contest the accuracy of the information on which their assessments are based.

Field Canvasses. The preponderance of information required in local government functions such as planning, assessment, and code enforcement is obtained by staff or contractor personnel through on-site, visual inspections. These inspections provide information on the characteristics of land parcels and of improvements on those parcels. Some jurisdictions recently have experimented with video technologies as a means of obtaining a visual record of properties.

Other Sources. The real estate industry, particularly real estate multiple-listing services, brokers, and private fee appraisers, can be a source of cadastral data.

5.3.2 Designing Data-Collection Forms

Data-collection forms traditionally have been made of paper. Portable data-entry devices, which can be used to collect, edit, and update records, may make such forms obsolete in computerized land-record systems. Nonetheless, some of the principles of forms design remain.

Data-collection forms may serve several purposes, depending on the application in question and on the design of the record system. A major purpose is to serve as either a temporary or semipermanent repository for information collected in the field. In a fully computerized cadastral record system, the useful life of a data-collection form ends when the data have been entered in and accepted by the computer. In a manual or partly computerized system, a form may serve as the official record itself, in which case its useful life will be indefinite, lasting as long as the property exists or until a new system is implemented. In manual or partly computerized systems, data-collection forms also serve as repositories of information about events and transactions and about administrative processes and decisions. In assessment, for example, manual data-collection forms, which are commonly known as property-record cards, also would contain information about sales, building permits, and assessment appeals and would document appraisal calculations and value conclusions (see Exhibit 5.1).

Data-collection forms also serve such purposes as facilitating the collection of property-characteristics data in the field, the conversion of such data into computer-readable form, and the making of those appraisal calculations done by hand.

Where the land-data registers are computerized, the design of a data-collection form is not likely to matter to any agency other than the one that collects and enters the data. Other users normally would want the data only in machine-readable form. However, where the shared data files are manual, the design of the forms will need to be a compromise among the needs of the several users.

In the design of forms for computerized systems, roughly equal consideration should be given to facilitating field operations and data entry; it is not necessary to provide for manual calculations or to be concerned about the format of reports, since the computer can reformat the data in any convenient way. However, consideration should be given to using cadastral record reports such as an appraiser's worksheet as a turnaround document for updating an existing record (Exhibit 5.2). If the system is a manual one, the compromise should consider ease of collection, ease of calculation, and ease of reading.

Other format and design issues include the size of the form, the weight of paper, and whether the form should be a single sheet, an envelope, or a folder (the latter two types being considered in manual systems in which supplementary documents may be part of a record). It is generally desirable to employ several specialized property-record forms rather than one all-purpose form. For example, an assessor might want separate forms for single-family dwellings, income-producing properties, industrial properties, agricultural properties, and so on. However, the forms should all have a similar format.

Data-collection forms should be designed to encourage accurate, complete, and consistent data. These objectives can be achieved by having variable labels that are clear, including variable numbers with each variable; having coding categories that are labeled, are exhaustive, and are mutually exclusive; providing sufficient space for recording numerical data; requiring a positive response for all variables, so that a blank means that the variable was overlooked and not that the property does not have the characteristic; and maximizing the use of checks or circles to ease and speed the recording of data.

CITY OF BATTLE CREEK-ASSESSOR'S OFFICE
RESIDENTIAL ASSESSMENT RECORD

MAP NO.   ASSESSMENT NO.

DATE TRAN.	RECORD OF OWNERSHIP	LIB.-PAGE	SALE

IMPROVEMENTS
- GRAVEL
- PAVED
- CURB
- SIDEWALK
- WATER
- SEWER
- ELECTRIC
- GAS

LAND
- LEVEL
- ROLLING
- LOW
- HIGH
- LANDSCAPED
- SWAMP

ZONING

SKETCH

PICTURE AREA

BUILDING or ALTERATION PERMIT

	DATE	AMOUNT
		$

LAND VALUE PLUS IMPROVEMENTS COMPUTATION

LOT SIZE	DEPTH FACTOR	EQUIVALENT FRONTAGE	RATE	TRUE CASH VALUE
			$	$

LAND IMPROVEMENT

	VALUE NEW	% COND.
WELL & SEPTIC TANK		
PAVED DRIVE		
FENCE & LANDSCAPING		

TOTAL LAND PLUS IMPROVEMENTS	$
TOTAL BUILDINGS	$
TOTAL TRUE CASH VALUE	$

YEAR	ASSESSED VALUATION	BOARD OF REVIEW	TAX COMMISSION
	$	$	$

INTERVIEWED	BY	DATE	ESTIMATED BY

ORDER BY FORM NO. M293K   RESIDENTIAL ASSESSMENT
FROM: DOUBLEDAY BROS. & CO., KALAMAZOO, MICH. JOB #526273

FORM APPROVED BY STATE TAX COMMISSION

EXHIBIT 5.1   Residential Assessment Record

						RATE	COMB.	FLAT	CORR.
						+ −	+ −	+	−

CLASS			NO. OF RMS. 1ST.	NO. BED RMS.	1-FAM.				
			2nd B		APTS.				
BSMT.	FULL		NONE						
	CONC. FLR.		PARTN.						
	FINISH								
FD'N.	BLK.		CONC.						
FRAME	WOOD	ALUM.	BLK.	INSUL. or ASB.					
WALL FACING	WOOD		BLK.						
	BRICK								
	MISC.								
ROOF	COMP.	ROLL							
	HARD		LINO. OR ASP. TILE						
FLOOR	SOFT		PRINT						
	CONC.		CERM.						
INTERIOR WALL	PLASTER - DRI-WALL		MISC.						
	TILE								
TRIM	HARD	SOFT							
	SINK	AMH	LT.						
PLUMBING	DISPOSAL	DISH WASHER	SHWR. STALL						
	STOOL	LAVATORY	TUB						
	3 PC. BATH	3 PC. BATH WITH SHOWER HEAD							
HEAT & AIR COND.									
PORCHES									
MISC.	FIREPLACE	OVEN	HOOD FAN						
	BLT. IN RANGE	ATTIC FINISH	DORMER						
	INCINERATOR								

TOTAL CORRECTIONS
NET CORRECTIONS

STY.	SIZE	AREA	COMB. AREA	RATE	NET RATE	AMOUNT

FLAT CORRECTIONS − PLUS − MINUS
CURRENT COST − LOCAL MULTP.
BASE REPRODUCTION COST

GEN. CONDITION	EX.	VG.	GOOD	AVG.	FAIR	POOR	V.P.
YR. BUILT	PART BUILT		AMOUNT	DEPR. x COND.	% DEP'R.		VALUE

DEPRECIATED REPRODUCTION COST
ENH. OR DET. INF.
GARAGE OR CARPORT - DEP. REP. COST
ECONOMIC CONDITION FACTOR
APPRAISED BUILDINGS VALUE

		BASE COST
AREA		
WATER		
INT. WALLS		
COMM. WALL		
DRIVEWAY		
TOTAL BASE COST		

GARAGE | CLASS
FDN. | SIDING
SERVICE DOOR | WIRING
RFG. | DOORS
FLOOR | DRAIN
CURRENT COST MULTP.
YEAR BUILT | DEP'R.
NOTES:

EAVETROUGH
STORMS
SCREENS
AWNINGS
THERMO PANE
INSULATION
EAVE OVERHANG

EXHIBIT 5.1 *Continued*

APPRAISAL WORKSHEET

PROP. CODE:				UPDATE ONLY	000 LAST UP DATE	/ /	BUILDING PERMITS:	
MAP NO.	BLK:	LOT:					DATE	NUMBER
PROP. ADDR.					NO. DWELLING UNITS		/ /	
OWNER:							/ /	
NBHD.	STRUCTURE CODE:			FIELD BOOK NO.		LAND USE CODE	/ /	

COST DATA

YR. BLT.	001	**ATTIC:**		**BUILT-INS (cont.):**		**PORCHES (cont):**		**RENTAL DATA:**	
GRADE	002	NO. RMS.	050	HOOD/FAN	092	2-CODE	140	CODE	190
% GRADE	003	% FIN.	051	COMPACTOR	093	AREA	141	GRM	191
TYPE	004	L.S.	052	RANGE/OV.	094	QUALITY	142	MONTHLY RENT	
YR. REM.	005	INDICATOR	053	INTRCOM	095	3-CODE	143		192
EFF.AGE	006	**BASEMENT:**		VACUUM	096	AREA	144	**STRUCTURE:**	
DEPR. COND.	007	NO. RMS.	055	SEC. SYS.	097	QUALITY	145	FND TN.-1	195
L.S.	008	AREA	056	KIT. REMOD.	098	4-CODE	146	FND TN.-2	196
INDICATOR	009	L.S.	057	OTHER	099	AREA	147	RF. TYPE	197
OBSOLESCENCE:		INDICATOR	058	L.S.	100	QUALITY	148	RF. PITCH	198
FUNC.	011	**FIRE PLACES:**		INDICATOR	101	5-CODE	149	PL. WALL	199
LOC.	012	HOUSE OP.	060	**PLUMBING:**		AREA	150	DRYWALL	200
PRIMARY STRUCTURE:		BSMT. OP.	061	4 FIX	103	QUALITY	151	PAN.VEN.	201
CONST.	014	CHIMNEYS	062	3 FIX	104	L.S.	152	UNF.WALL	202
STY.	015	L.S.	063	2 FIX	105	INDICATOR	153	PINE FL.	203
AREA	016	INDICATOR	064	EX. W.C.	106	**MISC. IMPROVEMENTS:**		HRDWD.FL.	204
CONST. II	017	ENERGY SRC.	066	EX. SINK	107	1-CODE	155	TILE FL.	205
CONST. III	018	L.S.	067	ROUGH-IN	108	AREA	156	CARPET/ SUB. FL.	206
ADDITIONS:		**ROOF MATERIAL:**		L.S.	109	GRADE	157		207
NUMBER	020	MATL. 1	068	INDICATOR	110	% DEPR.	158	BSMT. ENT.	208
CONST. 2	021	MATL. 2	069	**GARAGES:**		2-CODE	159	INSULATION	209
STY. 2	022	L.S.	070	1-CODE	115	AREA	160	PLB.BSMT.	210
S.F. 2	023	**BATH TILE:**		AREA	116	RATE	161	PLB.1STFL.	211
CONST. 3	024	FL. & W.	072	RATE		GRADE	162	BDR.BSMT.	212
STY. 3	025	FL. & SH.	073	GRADE	117	% DEPR.	163	BDR.1STFL.	213
S.F. 3	026	FL. ONLY	074	% DEPR.	118	3-CODE	164	BDR.2NDFL.	214
CONST. 4	027	L.S.	075	2-CODE	119	AREA	165	TOT. BDRS.	215
STY. 4	028	**HEATING-A/C:**		AREA	120	RATE	166	MET. WDW.	216
S.F. 4	029	H.A. FORCE	077	RATE		GRADE	167	STORM WDW.	217
CONST. 5	030	H.A. GRAV.	078	GRADE	121	% DEPR.	168	**EXCEPTIONS:**	
STY. 5	031	H. WATER	079	% DEPR.	122	4-CODE	169	UNIQUE PROP.	220
S.F. 5	032	FL. FURN.	080	3-CODE	123	AREA	170	POOR PLAN	221
PERI- METER	034	RADIANT	081	AREA	124	RATE	171	ACTS-NAT.	222
LIV AREA	035	BASE BD.	082	RATE		GRADE	172	UNFIN. VAL.	223
DORMERS:		HEAT PUMP	083	GRADE	125	% DEPR.	173	**STREET:**	
NO. FRT.	037	WALL UNIT	084	% DEPR.	126	L.S.	174	DEDICATED	225
% FRT.	038	NO HEAT	085	L.S.	127	INDICATOR	175	UNIMPR.	226
NO. SIDE	039	CEN. A/C	086	INDICATOR	128	MISC. COST	176	CURB	227
% SIDE	040	L.S.	087	**PORCHES:**				GUTTERS	228
NO. REAR	041	INDICATOR	088	1-CODE	135		180	SIDEWK.	229
% REAR	042	**BUILT-INS:**		AREA	136	**PROBLEM CODE:**		X-TRAF.	230
L.S.	043	DISHWASHER	090	QUALITY	137	DATA NO.	182	NON-THRU	231
INDICATOR	044	DISPOSAL	091			DATA NO.	183	CULDESAC	232

EXHIBIT 5.2   Example of Appraiser's Worksheet

			LOT-ZONING DATA:		LAND COMPUTATION:		
DECL. VALUE	PUR	% COMPLETE	LOT AREA		CODE		330
			WIDTH	DEPTH	UNIT PRICE •		331
		250	EXCESS AREA		LOC. ADJ.		332
		251	ZONING	MLT. ZONING	SIZE ADJ.		333
		252			PHYS. ADJ.		334

OTHER APPRAISAL DATA

UTILITIES:		INFL. (cont.):		SHAPE (cont.):		CONV. (cont.):		LAND COMPUTATION (cont.)	
								OUTSIDE INF.	335
ST. DRNS.	235	PRKWY	257	REAR	275	COMM.	294	ADJ. UNIT PRICE	
WELLS	236	HIWAY	258	DBL.FRT.	276	RELIG.	295	NO. UNITS	336
SEPTIC	237	INDUS.	259	TERRACE	277	SCHLS.	296	LAND VALUE	
LOCATION:		COM.	260	TOPOGRAPHY:		PRK-OFF	297	L.S.	337
WTRFRT.	240	R.R.	261	LEVEL	280	PRK-ON	298		
ADJ.PRK.	241	CEM.	262	AB. ST.	281	METRO	299	SUMMARY LS: COST:	
ADJ.GOLF	242	NOISE	263	BLW. ST.	282	NEIGHBORHOOD:		BASE VALUE	
VIEW	243	ODORS	264	ROLLING	283	IMPROV.	305	ATTIC	
PRIV.	244	NUIS.	265	STEEP	284	DECL.	306	BASEMENT	
RESTR.	245	LOT	266	FLD. PL.	285	STATIC	307	DORMERS	
ACCESS	246	SHAPE:		WOODS	286	PLN.COM.	308	FIREPLACES	
HIST.	247	RECTNG.	270	CONVENIENCES:		ARC.CON.	309	ROOF MATL.	
GL. PATH	248	TRIANG.	271	BUS	290	HOMOG.	310	BATH TILE	
INFLUENCES:		TRPZ.	272	SHOPS	291			ENERGY SOURCE	
CORNER	255	CURVED	273	PUB.SER.	292			HEAT-A/C	
ALLEY	256	L-SHP.	274	SOC.SER.	293			BUILT-INS	
REMARKS:							315	PLUMBING	

BUILDING DIMENSIONS:

LAND COMPUTATION / SUMMARY (cont.)
PORCHES
GARAGES
RCN TOTAL
DEPRECIATION   %
FUNC. OBS.   LOC.OBS.
RCLND TOTAL
IMPROVEMENTS TOT.
RCLND + IMPROV.
LCF.
ADJ. COST VALUE
LAND VALUE
MISC. COST
TOTAL COST VALUE

VALUATION SUMMARY:

LAST ASSESSED VALUE		
SALE DATE	/ /	
SALE PRICE		
COST VALUE		
MRA VALUE		
INCOME VALUE		
CORRELATION VALUE		
FINAL VALUE		
APPRAISER VALUE		350
DATE	/ /	351
I.D.	352 SOURCE	353

EXHIBIT 5.2   *Continued*

APPRAISAL WORKSHEET - DEPARTMENT OF REAL ESTATE ASSESSMENTS - CITY OF ALEXANDRIA, VIRGINIA

INFORMATION SOURCE
1 Owner
2 Neighbor
3 Maid
4 Child
5 Tenant
6 Estimated
7 Other

LAND COMPUTATION
1 Site Value
2 Square Foot
3 Front Foot
4 Acres
5 Effective Front Foot
6 Other

FOUNDATION
1 Concrete
2 Tile
3 Concrete Slab
4 Stone
5 Piers
6 Brick
7 Concrete Block

ROOF TYPE
1 Mansard
2 Gambrel
3 Flat
4 Gable
5 Hip
6 Expansion
7 Combination

ROOF PITCH
1 Flat
2 Low
3 Medium
4 High

BASEMENT ENTRANCE
0 None
1 Walk-Out
2 Semi Walk-Out
3 Outside Entrance

PORCH TYPE
01 Concrete Patio
02 " " (with cover)
03 Frame Stoop
04 " " (with cover)
05 Concrete Slab (3" on cinder)
06 " " " (with cover)
07 Concrete Stoop
08 " " (with cover)
09 Brick Patio on Concrete
10 " " " " (with cover)
11 Raised Wood Deck
12 Raised Wood Deck (with cover)
13 Flagstone in Sand
14 " " (with cover)
15 Brick Ornamental on Concrete
16 " " " " (with cover)
17 Flagstone on Concrete
18 " " (with cover)
19 Concrete Slab (5")
20 Concrete Slab (5") (with cover)
21 Pre-Engineered Cover
22 Open Porch with Cover (1 story)
23 Open Porch with Cover (2 story)
24 Screened Porch with Cover (1 story)
25 " " " (2 story)
26 Glass Enclosed Porch (1 story)
27 " " " (2 story)
28 Frame Enclosed Porch (1 story)
29 " " " (2 story)
30 Brick Enclosed Porch (1 story)
31 " " " (2 story)
32 Colonial Porch
33 Porch W/Basement-Frame (1 story)
34 Porch W/Basement-Frame (2 story)

PORCH TYPE (cont.)
35 Porch W/Basement-Encl.-Frame (1 story)
36 Porch W/Basement-Encl.-Frame (2 story)
37 Porch W/Basement-Brick (1 story)
38 " " " (2 story)
39 Porch W/Basement-Brick Encl. (1 story)
40 " " " " (2 story)
41 Porch Screened W/Basement-Brick (1 story)
42 Porch Screened W/Basement-Frame (1 story)

PORCH QUALITY
1 Below Average
2 Average
3 Above Average
4 Excellent
5 Mansion

MISC. IMPROVEMENT
1 Swimming Pool (Reinforced Concrete)
2 Miscellaneous Bldgs.
3 Tennis Court
4 Bath House
5 Greenhouse
6 Swimming Pool (Gunite-Vinyl Lined)
7 Swimming Pool (Welded Metal-Coated)
8 Swimming Pool (Fiberglass)

GARAGE
1 Attached Frame
2 Attached Brick
3 Attached Stone
4 Detached Frame
5 Detached Brick
6 Detached Stone
7 Basement
8 Carport
9 Built-In

ENERGY SOURCE
1 Electric
2 Gas
3 Oil
4 Coal
5 Solar
6 Other

Roof Material
1 Tile
2 Shakes
3 Metal
4 Roll
5 Asphalt Shingle
6 Asbestos
7 Built Up
8 Slate

DEPRECIATION CONDITION
1 Excellent
2 Good
3 Average
4 Fair
5 Poor

HOUSE GRADE
1 Seasonal
2 Economy
3 Average
4 Good
5 Expensive

HOUSE TYPE
01 Standard Unit
02 2 Family Side by Side
03 3 Family Side by Side
04 4 Family Side by Side
05 Row House-End Unit
06 Row House-Center Unit
07 Split Level
08 2 Family Stacked
09 3 Family Stacked
10 4 Family Stacked
11 A Frame
12 Pre-Eng. (Pre-Fab)
13 Mobile
14 Split Foyer
15 Bi-Level

HOUSE CONSTRUCTION
01 Wood Frame
02 Brick Veneer
03 Stone Veneer
04 Solid Brick
05 Solid Stone
06 Stucco Wood Frame
07 Stucco Masonry
08 Concrete Block
09 Tile
10 Metal
11 Wood & Brick
12 Wood & Stone
13 Siding on Sheathing
14 Single Siding
15 Wood Shingles
16 Composite Shingles
17 Aluminum Siding
18 Clapboard
19 Asbestos Shingle
20 Brick Tex
21 Perma Stone

STORY HEIGHT
100- 1 W/Basement-No Attic
120- 1¼ "
150- 1½ "
170- 1¾ "
200- 2 "
220- 2¼ "
250- 2½ "
270- 2¾ "
300- 3 "
320- 3¼ "
350- 3½ "
370- 3¾ "
400- 4 "
105- 1 W/No Basement-No Attic
125- 1¼ "
155- 1½ "
175- 1¾ "
205- 2 "
225- 2¼ "
255- 2½ "
275- 2¾ "
305- 3 "
325- 3¼ "
355- 3½ "
375- 3¾ "
405- 4 "
1008 1 W/Basement-Attic
2008 2 "
3008 3 "
4008 4 "
1058 1 W/No Basement-W/Attic
2058 2 "
3058 3 "
4058 4 "
10— 1 W/No Basement-No Attic
        (Post or Pier Foundation)
12— 1¼ "
15— 1½ "
17— 1¾ "
20— 2 "
108- 1 W/No Basement, W/Attic
        (Post or Pier Foundation)
208- 2 "

BUILDING PERMIT PURPOSE
1 Remodel
2 Addition
3 Renovation
4 New House
5 Air Condition
6 Plumbing
7 Demolish Building
8 Moved Building

EXHIBIT 5.2 *Continued*

5.3.3 Designing Data-Collection Manuals

A data-collection manual is an important element in a cadastral data-collection program. The objectives of a coding manual are to expand knowledge about the property characteristics being described, to achieve accuracy and consistency in describing and coding property characteristics, and to speed the data-collection effort. These objectives are achieved by describing and explaining the content of data-collection forms, with explanations of purposes, definitions, and instructions for each data item.

The purpose of an item should be given in cases where its intent is not obvious. Explanations of the purpose of collecting various items can assist data collectors to make correct decisions when confronted with unusual situations in the field.

Definitions of terms provide another crucial control on observations. They provide checks in several respects. They define the limits of the observations and are designed to elicit precise observations within the specified limit. They ensure that each data collector defines terms and evaluates items within the same frame of reference as every other data collector. They are often intentionally rigid so that there will be little opportunity for individual interpretation. A difficult aspect of the formulation of a good definition is to provide rules that, to the greatest extent possible, guarantee consistency and simultaneously allow enough flexibility to accommodate unexpected circumstances. Therefore, subrules sometimes must be built into definitions to accommodate special exceptions. Definitions should also be designed for internal and external consistency. That is, they should be considered as elements of an overall description scheme and never as isolated from or independent of one another. Finally, definitions must work for the data collectors who will use them. If the definitions are too simple, data collectors will become uncertain and confused. If the definitions are too detailed, they become intimidating, cumbersome, and unworkable. A delicate balance is necessary in designing definitions that yield good data. Often a picture or a sketch is the simplest way to define a characteristic.

Instructions are used where difficulty is anticipated in describing an item because detailed, specific observation is required. Each step entailed in describing the item should be specified in these cases. In addition, separate guidelines should be provided in cases where exceptional circumstances are anticipated.

5.3.4 Editing and Auditing Cadastral Records

Data edit and audit procedures are an integral part of an overall effort to ensure that cadastral records are complete and accurate. Data edit and audit procedures provide notice of errors in the data and warnings about possible errors or unusual situations. They can be done manually or computerized, although computerized edits generally are more thorough and cost-effective. Computer-generated edit reports can indicate whether the condition that has been detected is an error or is a warning and can briefly describe the condition. Edit routines should check the following:

Missing Data. An error message should be issued each time missing data are detected. Data-coding procedures should minimize value blanks in the data.

Valid Characters. An error message should be issued if invalid characters are detected, that is, the routine should ensure that only numerical characters are used in numerical fields and alphabetical characters in alphabetical fields.

Valid Codes. An error message should be issued each time there is an invalid code. For example, a 3 appearing in a dummy-variable field where only 0 or 1 is valid is an error. Similar checks should be made of use codes, map numbers, and the like.

Normal Ranges. A warning message should be issued each time a numerical value falls outside a prespecified normal range. For example, an assessor might specify that the normal range in total floor area of residences in a neighborhood is between 800 and 3000 square feet. A house with a total floor area outside this range would trigger a warning message. All warning messages should be checked out, and data errors should be corrected. It is desirable for edit routines to have a way of flagging valid exceptions to normal ranges. Suppose, for example, that a house in the neighborhood discussed above had a total floor area of 3200 square feet. A flag acknowledging this fact would save the effort of checking out total floor area each time the edit routine was run. Judgment must be exercised in setting normal ranges. Tight limits will result in unnecessary warning messages, whereas loose limits will result in too few messages.

Data Consistency. Error and warning messages can be issued whenever an inconsistent or illogical relationship exists between the recorded data on two or more property characteristics. For example, a count of bedrooms in excess of the total number of rooms should result in the issuance of an error message. An assessor might decide that the number of bathrooms is normally less than the number of bedrooms, and bathroom counts equal to or exceeding bedroom counts would cause a warning message to be issued. Another consistency edit might be to divide the total floor area of a building by the number of rooms and compare the resulting average area per room with a specified normal range of area per room ratios. Deviations from this normal range might indicate errors in room counts or errors in area measurements. The range of consistency edits is limited only by the ingenuity of the editor and the resources available to investigate possible errors.

Check Digits. Parcel identifiers and other numerical quantities can have a check digit assigned to them that helps prevent transpositions and other data-entry errors.

Edit routines should perform all edits on a property record before moving to the next record. Some edit routines move to the next record after the first error or warning condition has been detected, leaving the detection of other error-warning conditions to subsequent edit runs. Such routines are inefficient and demoralizing to the staff assigned to checking error and warning messages.

In some cases it is desirable to reinspect a sample of properties to ensure that information is being coded accurately. This is an expensive quality check and should be reserved for checking the work of new, temporary data collectors and trainee appraisers and for verifying the quality of the work of contract data collectors.

At every stage of the data-collection program, data completeness and accuracy should be stressed. Errors and omissions are expensive to correct.

5.3.5 Data Maintenance

Property characteristics are always changing, and, if an effort is not made to keep the data on property characteristics up to date, the resources expended in collecting the original data are soon wasted. There are basically two approaches to maintaining property characteristics data: building-permit monitoring and periodic reinspections.

As previously mentioned, building permits are used to alert assessors and others to changes in properties. When a building permit is received by an assessor's office and the permit is for an assessable construction activity, the property's property record form should be pulled or flagged so that the construction activity can be monitored. After the data collector has determined that construction activity has stopped or the permit is no longer in force, the record should be returned to the property-record file or the flag removed.

No matter how good a building-permit reporting and monitoring system is, undetected changes in properties will occur continuously. Cadastral record managers should therefore periodically reinspect all properties in order to verify and update the information on hand for each property. Annual visits are optimal from an appraisal accuracy standpoint, but visits of that frequency may not be administratively feasible. Visits should at least be scheduled in conjunction with reappraisals. It is important to note that the chief function of these inspections is to verify rather than to collect information. Therefore, a drive-by inspection, during which the property and its record are visually compared, is often sufficient. Two-person teams of appraisers, in which one drives and the other handles records, can review and verify several hundred records per day. Visits may be supplemented with information obtained from taxpayer returns and from an examination of aerial photographs. Changes indicated by these sources should be verified in the field. Information supplied by taxpayers during assessment review and appeal proceedings can likewise alert assessors to inaccurate or out-of-date information on property characteristics.

5.4 SYSTEM DESIGN AND DEVELOPMENT PROCEDURES

The managerial, technical, and communication skills of system designers, users, and local government managers are challenged in the development of a land-records system based on the multipurpose cadastre. A hallmark of a successful system-development activity is careful attention to each step in the system-design, -development, and -implementation process, which is sometimes called the system "life cycle" (Young, 1980a). As previously mentioned, defining user requirements (see Section 5.1) is one of the early steps in system design and development. Other steps in the process include planning, analysis, design, development, and implementation. These subjects are briefly reviewed below. The discussion deals generally with application-system design, since record-system design generally is an integral part of an application-system design effort. More comprehensive treatments of system-design procedures are found in American Public Works Association Research Foundation (1981a), Auerbach Publishers (1980b), Donaldson (1978), Fife (1977), and Giles (1974).

5.4.1 System Planning

Effective project planning is crucial to the successful development and use of a multipurpose cadastral record system. The plan identifies needs that the system is to meet, the steps to be taken in implementing the system, resource requirements, and timing considerations. The plan also serves as a project control tool and is used to measure progress. Necessarily, the plan should be in writing.

A major part of the planning effort is scheduling. Scheduling involves dividing the overall process into discrete tasks and subtasks, noting which tasks can be begun only after other tasks have been completed and which tasks can be performed simultaneously; estimating realistic production rates and available resource levels, particularly personnel, for each task; and depicting time and resource requirements for each task on a Gantt chart or some other form of bar chart. A refined schedule should be made during the system-development phase as soon as major features of the system are decided on. More sophisticated project planning and management techniques, such as critical-path management/program evaluation and review techniques (CPM/PERT), should be used if the project is large and well analyzed.

CPM/PERT analyses are built around network diagrams that show the sequences of tasks as well as a single estimate of each task's time requirement and cost per unit time (CPM) or three estimates of each task's time requirement such as optimistic, probable, or pessimistic (for PERT). These estimates are then analyzed by machine or by hand to identify those tasks that are "critical" as opposed to those tasks that can be prolonged. Total estimated time is then compared with available time. If available time is inadequate, the project may be "crashed," which means determining which tasks can be most economically accelerated in order to meet tighter time requirements. As the project progresses, the project manager notes time (and cost) milestones achieved and crashes the remaining tasks as necessary when delays are encountered.

Sometimes a feasibility study is conducted as part of the planning process (Young, 1980b). In many respects, a feasibility study is a preliminary system design effort, the purpose of which is to determine whether the system can be justified in terms of services, costs, or both.

5.4.2 System Analysis

The functions and responsibilities of the agencies that will be using the multipurpose cadastral record system must be carefully analyzed to determine the scope of the various applications and, therefore, the capabilities required of the system. System analysis should begin to identify the resources required in system design, system implementation, and system maintenance. The analysis should identify data requirements and should coordinate data-element definition and related requirements of the various users. An estimate of the benefits and costs associated with each application will help determine priorities among the various possible applications (King and Kraemer, 1980). Consideration should be given to satisfying expectations of early payoffs by developing the system in a modular fashion, if possible.

System analysis begins with interviews with key individuals in user agencies and groups. The purpose of these interviews is to determine functional responsibilities; information needs; analytical and decision-making processes; and sources, availability, and condition of existing data. An interview guide or an interview form should be prepared to ensure that all relevant lines of inquiry are followed with each interviewer. Copies of procedural manuals, forms, and reports that are used, processed, or prepared should be carefully reviewed. A helpful intermediate step in the analysis of information flows is to create two matrices, one showing data elements and their users and the other showing data elements and their sources. The interview notes and the matrices are then used to prepare an accurate description of information processing by the entities involved.

5.4.3 System Design

The system analysis is used as the basis for system-design activities. The first step in designing the system is to develop a rough concept of what the system is to be. Brainstorming sessions involving key users and technical personnel can be helpful in the conceptualization process. After decisions on the general system features are made, the design effort turns to more detailed concerns.

The system itself should be decomposed into tasks. A narrative description of each task should be prepared. The narrative would indicate whether a particular task or subtask was to be performed by human or by computer. For each computer task, a program solution should be prepared to guide the programmers. A data dictionary also should be prepared.

There are several useful new programming techniques that are oriented toward the human side of computer use. The techniques have different labels applied to them—for example, "human engineering," "functional design," and "improved programming technologies." Components of the techniques, also identified by buzz words, include top-down program development, hierarchy plus input-processing output (HIPO) as a design and documentation aid, structured programming, chief programmer teams, development support libraries, and structured walkthroughs.

These techniques offer a number of benefits. The data base is defined early, before programs are written to retrieve information. This helps to avoid problems that sometimes arise when different parts of a program are written by different programmers at different times—parts that later have to be meshed together. System and program documentation is written as the system design and program structure are developed. Documentation is, therefore, more complete, accurate, and useful. The documentation also is organized in hierarchical levels, usually by function, making it easier to locate specific components of the program structure. Program code is easily intelligible to other programmers, and programs are easier to modify. The techniques also necessitate regular communications among users, system-development personnel, and data-processing operations personnel. Taken together, these features of the improved programming techniques can increase the confidence of both programmers and users in the programs and, therefore, in the system.

5.4.4   System Development and Implementation

System-development activities take place concurrently with system-design activities. An early decision is whether the system is to be developed internally or acquired from some external source (see Section 5.5). No general recommendation can be made as to which alternative is preferable. On the one hand, complete reliance on internal development may result in system-design personnel redeveloping existing systems while ensuring that the system meets the specific needs of the locale in question. On the other hand, systems developed elsewhere are seldom, if ever, completely transferable and, if they were, probably would not meet all the requirements of the host users. In fact, there appears to be an inherent contradiction in designing systems that are at once integrated and transferable. Of course, system transfers can occur at several levels, ranging from system concepts down to specific program code, and system components usually can be transferred more easily than entire systems.

A related question is whether to use external technical assistance in the system-development process. Actual experience with the use of consultants is mixed. Consultants can be a source of expertise not available locally and also can augment the system-development work force on a temporary basis. Much depends on the client's ability to define clearly the work products expected from technical consultants as well as the delivery schedule. In addition, the contracting agency must have the capability of managing the contract and monitoring the contractor's performance.

With respect to project management in general, several observations and recommendations can be made. First, the project planning and management techniques and the programming technique mentioned earlier can be of assistance in system development. The steady flow of products resulting from such techniques makes monitoring the project easier. Moreover, parts of the system can be tested incrementally, in contrast to a massive testing effort at the end of the development phase. The techniques also give managers, system-design personnel, and users a clear picture of the system as it evolves, thereby making it easier to spot errors and omissions and to suggest modifications and improvements.

System-implementation activities revolve around making sure that the system performs the way it was designed to. The information that the system receives must be converted into a form that the system can use. The chief recommendations that can be made in this regard are (1) to use existing data if practicable, since data collection is very time-consuming and costly, and (2) to take all feasible steps to ensure that only accurate data are stored in the system.

User orientation is a major activity during the system-implementation phase. User manuals are prepared, and orientation and training sessions are held.

5.5 ACQUIRING COMPUTING CAPABILITIES

Having cadastral records in computer-readable form offers innumerable benefits and is now feasible even for the smallest counties. Computers can reduce the time spent on such mechanical processes as producing reports and documents, sorting records, and aggregating data. They can speed mathematical calculations and make possible statistical analyses that could not be feasibly done manually. Increasingly, computers are being used to produce maps and other graphic displays. In addition, computerized diagnostic checks can enhance the quality of data.

Quite naturally, cadastral record managers often are concerned with evaluating and acquiring or upgrading computer hardware, software, or both. The range of considerations that enter into the decision-making process is quite broad, and only major points can be touched on here. Managers should turn to experts for the additional experience they need. One way to keep abreast of developments in the field is to consult the *ACM Guide to Computing Literature* (Association for Computing Machinery, annual).

An early step in acquiring computing capabilities is to develop a general strategy (Donaldson, 1978). Unless one is constrained to use computing machinery currently on hand, software needs should be evaluated first, and hardware needs should be evaluated in the light of requirements imposed by the software and the amount of data to be manipulated.

A general issue is processing mode. Formerly most processing was done in a batch mode, which offered some processing efficiencies. However, the trend is toward on-line processing, in which files are updated on a record-by-record basis at the instant the user chooses. This is accomplished through terminals, and the system must be designed to accommodate a variety of inquiry, update, and other jobs being performed more or less simultaneously.

There are several alternative strategies to acquiring software: writing the programs that are needed using some combination of in-house and consulting personnel, attempting to adopt software that is in operation elsewhere, and purchasing packaged software. Guidance on choosing among these alternatives, particularly the latter, can be found in American Public Works Association Research Foundation (1981b), Brown and Stephenson (1981), Dekle (1981), *Planning* (American Planning Association, 1981), and Roberts (1980), among others. Currently, the first alternative may be the only alternative if the objective is to develop an integrated multipurpose cadastre. Sources of information on the management of software-development projects include Donaldson (1978), Fife (1977), and Metzger (1973). The feasibility of choosing the second alternative depends on the nature of software needs. If the goal is to develop an integrated multipurpose cadastre, the second alternative is not likely to be feasible because there are comparatively few models to draw from. If only a module or two are needed, the second alternative may be feasible. However, "borrowed" software may not be well documented, and support will be limited.

The basic approach to evaluating software packages is to establish a formal structure for comparing alternatives. While there is considerable variation in printed checklists (American Planning Association, 1981; Brown and Stephenson, 1981; Dekle, 1981), there are six major points of comparison: (1) how well a given package fulfills the user's requirements, with appropriate consideration of priorities among the requirements; (2) the history of a given package; (3) the strength and reliability of the vendor (the likelihood that the vendor will be able to provide support down the road); (4) software support (documentation and training); (5) hardware requirements (the processing efficiency of the package); and (6) short- and long-term costs.

Hardware should be evaluated in terms of such factors as storage capacity, processing speed, reliability, availability of service personnel, human engineering features, and expandability or upgradability. Computing capabilities can be acquired in several ways. The machinery can be purchased or leased. Machinery alternatives include large computers (mainframes), minicomputers, and microcomputers. Both Auerbach Publishers, Inc. (6560 N. Park Drive, Pennsauken, N.J. 08109) and Datapro Research Corporation (1805 Underwood Boulevard, Delran, N.J. 08075) publish guides to computers and peripherals that can be used in identifying suitable hardware. Other alternatives to acquiring computing capabilities include facilities-management contracts, joint-powers arrangements, regional computing authorities, the purchase of computing services from another government, and service bureau contracts (King, 1980). Cooperative computing endeavors, such as are inherent in multipurpose cadastres involve a number of management issues that will need to be addressed (Almy, 1979b; Bernard, 1979).

The managers of cadastral record systems have the same general responsibilities in acquiring computing machinery on services as they have in procuring any other major product or service; that is, they must protect the public interest while ensuring that all potential suppliers have an equal opportunity to bid. Local purchasing regulations and state procurement codes affect contracting procedures. The cadastral record manager's main concern is with specifying the work to be done and the capabilities that products must possess.

Usually the first step is to prepare a set of specifications and to issue a request for proposals (RFP) (Motto, 1980). The purpose of the specification is to define the scope of work, the standards of performance, and the respective responsibilities of the government and the contractor. They should be prepared before specific firms are considered. The RFP and the specification should be sent to several potential contractors.

The second step is to select a contractor. The amounts bid by potential contractors should not constitute the sole grounds on which the selection is made. Proposals also should be evaluated on the basis of the contractor's responsiveness to the project specifications, technical qualifications, and financial responsibility. The contract can then be awarded on the basis of the lowest among the bids received for comparable systems from qualified, responsible bidders (Matthews, 1980).

The final step is to monitor the work of the contractor carefully to ensure that work is proceeding according to contract specifications.

5.6 SECURITY AND CONFIDENTIALITY

Managers of cadastral record systems need to be concerned with record security and confidentiality as well as record completeness and accuracy. They need to devise procedures that fix responsibility for work (i.e., develop an audit trail), restrict access to some information in the records, prevent unauthorized changes to records, prevent loss of records, including catastrophic loss, and minimize the possibility of malfeasance, as well as prevent erroneous changes to records (see Section 5.3.4).

In manual cadastral record systems, security and confidentiality can be ensured by placing records in a secure place, restricting access to those records to authorized personnel, and keeping track of the people who have had access to records. A wider variety of security and confidentiality measures can be employed in computerized systems. In particular, access to certain key tables in an assessment system may be restricted to prevent unauthorized persons from changing valuations. Similarly, access to information about the financial affairs of individuals and business is often restricted because that type of information is viewed as being confidential. Transaction logs or journals may also be maintained, which provide an "audit trail" of each change, noting what the change was, when it was made, and who made it.

Most types of land information are considered to be public information and have not been involved in the privacy and security controversy that has surrounded the computerization of personal data about individuals. Nevertheless, concern that computerization will make it easier to compile information about the land ownerships of particular individuals or about land transactions in particular areas may raise some concern. In the fear that this type of information violates privacy, objections may be raised to implementation of cadastres. System developers should remind those objectors that no new information is being made public.

A distinction needs to be made. The cadastre does not conclude that someone is misusing land or reaping undeserved gains. Another person has to interpret the data and draw conclusions. The emphasis should be on the interpretation, not the system. Keep the system out of the controversy.

6

The Evolving
Land-Information
Environment

Land information is most useful if spatially referenced. One way of accomplishing this is to use land-ownership parcels as the unit of observations, as described in Chapter 5. This allows comparison among parcels or among aggregations of parcels. Alternatively, land information can be organized into other types of homogeneous units, which, like parcels, are observed as polygons having uniform character. Or an arbitrary grid can be imposed over the mapped data and land characteristics then attributed to each grid cell. First, we examine land-ownership parcel schemes for spatial referencing.

Among the wide range of Geographic-Information Systems (GIS) described in Section 1.4.4, the systems that attribute land data to the visual center of homogeneous polygons or grid cells are classified as recording land digitally in a *discrete* rather than a *continuous* manner. Discrete systems contain data on units of observation, say for parcels, city blocks, or homes, but the emphasis is on the units and not on how they relate to each other at their boundaries. Each unit is treated independently; the boundaries of the units are not described. On the other hand, continuous systems are digital maps that partition the land space with points, lines, and areas describing the spatial extent and juxtapositions of land parcels or other natural and cultural features that make up the landscape.

Parcel-related records organized by parcel index numbers constitute a discrete information system. The records are often processed without reference to a cadastral overlay, which locates the parcel boundaries. Discrete parcels serve as units of observation in the system. For limited purposes, discrete Land-Information Systems are effective and relatively easy to use.

Land-parcel data can be considered continuous rather than discrete when keyed to a map of parcel boundaries. The map locates the data over continuous space rather than just by discrete parcels. This continuous framework enables integration of land parcels with other continuous land data.

6.1 INTEGRATION OF DATA THROUGH SPATIAL REFERENCES

The geodetic framework of a cadastre provides for integration of land-ownership information with other information, if the latter is spatially registered according to the same geometric framework with some indication of positional accuracy. This ability to relate data spatially is a powerful tool in the analysis and management of land and activities. As Figure 3.1 illustrated, the overlaying of layers of data provides an ability to meet a wide range of diverse needs.

Manual overlay of map products meets many needs for mapped information but depends on human integration of the composited information. Manual overlaying is limited by the content choices of the various layers made by the map designers. Users can select layers to composite for their particular application, but any variation in scale must be accomplished by photographic enlargement or reduction, and variation of map content or data categories requires remapping.

In digital form, users have a choice of the themes or layers to composite and a choice as to detail and ranges in terms of the content. In addition, choice of scale is accomplished mathematically.

Base mapping according to national mapping standards is central to the multipurpose cadastre concept. This enables spatial registration of data layers. Spatial registration of the map graphics would not be an issue if locations of all land data were recorded as numerical field measurements, as they are for property boundary corners (see Section 4.2.1). However, because many users will continue to use aerial photo imagery as the source of their location data, especially for natural phenomena, determining their coincidence in space will continue to depend on accurate two- or three-dimensional plotting. Adherence to base-mapping standards avoids subsequent problems of separately mapped data not correctly relating spatially.

Poor-quality base mapping or attempts to collect spatial data from uncontrolled maps will create spatial registration problems. To a certain extent these problems can be anticipated and dealt with. This may mean visual and manual reconciliation of spatial inconsistencies on a single base map and then digitization of the manually reconciled spatial data. Alternatively, each data layer or theme can be digitized from whatever source maps are available, and then the data can be rotated, transformed, and scaled to fit. This may require considerable editing in terms of redigitizing of point and line data to achieve fitting. A major problem is "slivers and gaps" that occur (1) when common boundaries between polygons of a single layer are separately digitized; (2) when boundaries are common across themes, such as where a property boundary is a street right-of-way; or (3) where a political boundary is a natural feature.

A multipurpose cadastre offers relief from most of the spatial registration problems in that all data are reconciled to a common base before or as part of the digitizing process. Nevertheless, inexactness in data-capture technologies requires care in the capture of spatial data, so as to retain the spatial registration in digital form.

In sum, spatial registration provides greater flexibility in the digital data-capture process. Without spatial registration one is forced to compile separate data for the same base and then digitize or resolve discrepancies among different bases. With registration, coordinate data for different layers or themes can be related directly. One does not have to go from digital spatial data to a map product and back to a digital data base. Instead, the digital data from orthophoto map generation can be incorporated directly into the cadastre along with the coordinates from property descriptions.

6.2   APPLICATION OF GEOGRAPHIC-INFORMATION-SYSTEM CONCEPTS TO LAND INFORMATION

A digital base map and a cadastral overlay are the key elements of a multipurpose cadastre, which enable it to become the basis for a powerful Geographic-Information System. An information system consists of a data base, with the necessary input, storage, retrieval, and output technologies responsive to nonroutine queries. A GIS is a special case where the data base is a digital map or consists of observations on spatially referenced features or activities, which are definable in space as points, lines, or areas. A GIS manipulates these spatial data to retrieve data for queries and analyses (Dueker, 1979).

Digital maps generally are formatted as *either* vector or *grid* data. Vector data describe areal features as polygons and linear features as line segments, both composed of digitized points. Grid data partition land space into a regular lattice with location specified by address or row and column numbers. Vector-format digital maps are employed for engineering, utility, and tax map applications, while grid-format digital maps are employed for thematic mapping and resource-analysis applications.

Vector data in the form of polygon encoding of coordinates capture geometric shape and location of features. Vector-format data in the form of topological facts and metric location and shape create an even more explicit digital map. The elementary objects in two-dimensional topology are points, lines of any shape, and areas. The relations among them are the incidence of areas and points separated by lines. A topological data structure enables the construction of a consistent digital map that contains relations among features, such as to select areas bounded by specific lines or lines that end at specific points (White, 1982). Topology provides capability to edit vector data and ensure logical consistency. Topologically structured vector data are essential in the creation of large digital map files.

When producing line maps from selected layers and items, vector-format data are used most often. When performing analyses that require relating data across layers, it is usually more convenient to convert from vector format to grid format. The user can specify the appropriate resolutions for the quality of the data or the resolution requirements of the analysis. This establishes the grid cell size at which the vector-described areas are resolved, so that corresponding cells can be compared to determine the interrelationship of factors, say, for example, land use by soil type and by ownership type. The resultant overlay of these three factors enables area measurements and production of maps showing, say, vacant parcels of 40 to 160 acres with class II soil that have changed hands in the last 2 years.

The polygon-to-polygon comparison of one polygon set to another polygon set is computationally cumbersome—calculating intersections and keeping track of new polygons. Gridding or rastering the polygon sets and comparing and tabulating corresponding cells is computationally more efficient than direct overlaying of polygons. The gridding of polygons prior to overlaying avoids the problems of slivers and gaps that result from imprecision in digitizing layers separately. A similar but even more efficient technique uses horizontal scan lines that intersect the polygons to perform the overlay.

6.3 EXCHANGES OF DATA BETWEEN LAND-INFORMATION SYSTEMS

Implementation of a multipurpose cadastre greatly simplifies the spatial collation of data. It not only eliminates the inherent duplication of mapping among utilities and governmental jurisdictions but also facilitates assembling composites, for example, of sewer, water, power, and gas lines or of ownership, floodplain, and agricultural land.

Notwithstanding these potentials, there exists in the short term a need to relate data compiled on different base maps and with inconsistent control. This might occur when land-resource data, say floodplains or agricultural land compiled on 1:24,000 U.S. Geological Survey 7½-min quadrangle maps, are related to land-ownership data compiled at 1:1000 to 1:5000 scale. The man-made boundaries can be defined with precision without unreasonable expense. It is difficult to define most natural boundaries, as a practical matter, with the same level of precision.

Care must be exercised in relating data from different sources. Table 6.1 illustrates the problems with transferring data between the different map scales normally used for cadastre and for resource thematic systems. This requires judicious choice of resolution of grid cell size so as not to lead to false accuracy assumptions or inaccurate allocations. This problem is a direct result of an order-of-magnitude difference in the scales at which the data were compiled. Resource thematic data such as soils and floodplain boundaries, are normally compiled at map scales between 1:10,000

TABLE 6.1   Problems with Exchange of Data between the Cadastre and Resource Thematic Systems

	Applications	
	Cadastral Data Systems	Resource Thematic Data Systems
Scale	1:1000	1:10,000
Natural Boundaries (soils, floodplains)	Enlargement implies ◄——— Compilation scale unwarranted accuracy	
Man-made Boundaries (parcels, political jurisdictions, building locations)	Compilation scale ———► Reduction produces unnecessary detail	

and 1:100,000. Transferring these already imprecise boundaries, whether by hand or by computer, to a cadastral mapping scale (1:1000 to 1:5000) implies a higher accuracy than warranted, which may create erroneous information relating to specific parcels of land. The solution to this problem lies in compiling the resource data at the scale needed for cadastral applications—an expensive approach. Otherwise, the boundaries must be drawn implying uncertainty, i.e., an imprecise, wide line.

A similar problem exists in transferring exact boundary data from the cadastre for use in resource thematic systems for environmental applications. This problem has two components. One problem component is that the process of converting coordinate location from one scale or projection to another may place the transferred data in an erroneous position with respect to data already compiled on the resource thematic map. Second, the volume and detail at the cadastre scale may not be needed for resource thematic applications. Smoothing or aggregation or both of the detailed cadastre data may be warranted.

A primary mode of analysis for environmental data is to overlay one factor with one or more other factors, such as soil type with land-ownership parcels. Choice of a cell size or scan-line interval is the means by which an overlay that relates inexact boundaries to exact boundaries can be accomplished. When overlaying imprecise boundaries, the analyst should select a large cell size, which will reduce the chance that a cell will be assigned to a polygon incorrectly. As a rule of thumb, the cell size or scan-line interval should be greater or equal to the width of the most imprecise line. Figure 6.1 illustrates the problem of ambiguity of cell assignment when the width of the imprecise line is greater than the width of one cell. If too many cell centers fall within the width of the line, there is uncertainty as to their correct assignment.

If the need for resource analysis precedes the development of control and base mapping for cadastral purposes, the option of developing a separate and less-accurate

Determination of Cell Size

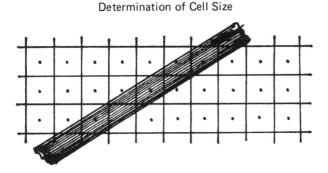

FIGURE 6.1   Illustration of the problem of relating small cells to a wide line.

resource thematic system must be considered. Resource thematic systems at a coarse scale take less time to develop than does a detailed cadastre. Resource problems often will not await the development of the cadastre on which to build an accurate resource thematic system. Further, many of the resource thematic applications may not require the map scale/resolution inherent in the cadastre, which may yield a volume of detail that is overwhelming. Experience in urban transportation planning shows that the land-parcel data sets provided more detail than was needed for metropolitan transportation planning. The data were immediately aggregated to a higher level—the traffic zone level—for analysis. The sizing and location of arterial highway facilities could be performed better with data aggregated to areas at least several city blocks in size. The same may be true for most resource planning and management analysis.

In the short run, the problem of exchanging data between the cadastre and other Land-Information Systems is one of spatially adjusting the data by using the computer or of recompiling the other land data on the cadastre base maps. This should be only a short-run problem, because once the base map from the cadastre is available it can be rescaled as the base for all new resource inventories, which then can be fit directly to the cadastral data. Further, the denser geodetic reference framework will also be available for new issues of map products of the U.S. Geological Survey—often used as the standard base for resource thematic mapping—and these products will be consistent with the base map of the cadastre.

7

Organization and Budget for a Multipurpose Cadastre

The Committee on Geodesy report (1980) explained the importance of building a system of cadastral records to serve each locality in the United States.

We reiterate the recommendations in the Committee on Geodesy (1980) report for actions to be taken by local, state, and federal governments to organize the development of a multipurpose cadastre for each locality in the United States. In particular, we recommend that states enact legislation to ensure the compatibility of county and local records with the multipurpose cadastre.

Chapters 2 through 6 of this report have described procedures and standards for upgrading each of the components of a multipurpose cadastre and ensuring that they will fit together into a coordinated system, with constant updating thereafter. The activities that are served by a multipurpose cadastre, whether governmental or private, occur mostly in the same locality as the properties that are described. The governmental functions listed in Chapter 5 as either sources or users of land data typically are parts of county government.

We recommend that the cadastre be organized as a function of county government in most localities in the United States, the exceptions being where municipalities assume responsibility for all or major parts of the governmental functions that are users of the records.

7.1 FACTORS THAT SHAPE THE LOCAL CADASTRAL RECORD SYSTEM

The arrangements for maintaining a multipurpose cadastre among the offices of a county or municipal government will vary as much as does the structure of those governments. The tasks of developing and operating a new level of technical operations will be assigned differently among the agencies of the local government, depending on such factors as the scope of authority provided in state-enabling legislation, the year in which the new system is designed, the resources available at that time to each of the several local agencies involved, and various other factors described in the following paragraphs. The differences in organizations should not prevent the local offices from meeting statewide standards for the contents and the operations of the multipurpose cadastre, as long as the latter are known and understood.

7.1.1 The Issue of Centralization: Operations versus Control

For efficient management of an information and record-keeping system, the management personnel of the user agencies must be involved. For a multipurpose cadastre, this means building links among three or four of the separate divisions of county or municipal government listed in Section 5.1. Perhaps the simplest arrangement for involving all these agencies is to assign the operation of the cadastre to the central administrative office, close to the chief executive. Most of the cadastres of West Germany, for example, are operated by the state finance ministries, in part owing to their importance in the taxation system.

In local governments in the United States the need for a broader and more efficient system of land records is being felt more immediately by the management of the user agencies than it is by the general public. *Prior to the assignment of the development of the cadastre to a central administrative office and the gathering of the necessary public and financial support thereof, a consortium of county-operating agencies should initiate design of the cadastre elements.*

Where an information system depends on cooperation among several suppliers of information to adhere to common standards and procedures, then it is helpful if each participant carries responsibilities for maintaining parts of the system that are in the same relative proportion as his or her need for the overall common results. The four most likely user agencies to be served by a successful local land-data system, according to a poll of expert opinion in February 1982, are the assessor, the planning department, the deed recorder, and the county or city engineer (Lincoln Institute of Land Policy, 1982). Other important users are listed in Section 5.1.

As recently as the early 1970's, the economics of data processing appeared to dictate that information systems be integrated for each district and operated from a central location. With the technology of microprocessing, this is no longer a given.

However, the decision as to how best to distribute an information system among several major participants takes on several dimensions.

The technology of "distributed data processing" provides only one of many options available and is not likely to be the most appropriate for many counties. Such a system is understood by data processors to have only the equipment decentralized. Control of computer operations remains tightly integrated, and much of the control itself is automated. This involves a much higher level of automation and control than will be appropriate for most county cadastres, many of which will continue to rely heavily on manual procedures for years to come.

The possibility of decentralizing the control of the use of the components of the data system among the participating agencies is a more relevant issue for most counties. The traditional model for computerized files is to have a common set of user priorities with which all must comply, administered by a data-processing agency, typically using a central mainframe computer. At the other extreme, the automation of the assessor's files in Toledo-Lucas County, Ohio, is proceeding with several microcomputers that cost a few thousand dollars each, with one staff person assigned to manage all the assessor's records for each of the several sectors of the city using his or her dedicated microcomputer. Data output can still be provided quickly to any of a variety of users, in machine-readable form if needed. However, the process of obtaining it obviously is highly decentralized.

Nevertheless, the selection of *software*, and control of *formats of data* in the files, is best done centrally among the participants in the cadastral record system. This will be essential if data are to be shared among the participants in an on-line mode. Otherwise, if each of the participants controls his or her own software, then exchanges of data can only be done in a batch mode, for example, by generating tapes of the desired data periodically for each of the other users. The latter would be necessary to allow for reformatting of the data for the user's different software, when it is transferred from the common source file onto a tape for the specific user.

Whatever the configuration of the system, the data in each independent file must contain a common spatial reference. Too much decentralization of the control of *data definitions and quality standards* can be seen as one of the major problems of local cadastral records as they exist today.

We recommend that a central office in the government of each county (or major city, where appropriate) be assigned the responsibility for managing the development of the systems of maps and files that will comprise the multipurpose cadastre for that locality and for compiling the common set of standards for definitions of data elements, file formats, accuracy, frequency of updating, and completeness of the records.

Leadership in the area of Land-Information Systems should be assigned either to a new agency designed for that purpose or to an existing agency that is capable of providing it. The central management should be seen as a permanent function to enforce continued adherence to the standards of data quality and to provide leadership in working out future stages of improvements in the system. These roles will be especially important where the development work itself is dispersed among user agencies.

In localities where the four major user agencies listed above are split between the county and municipality, for example, the issues of organizing the multipurpose cadastre become rather complex and probably will require a political resolution at the state level. The locations of the property-assessment function can be expected to set the overall pattern of responsibilities for the development of the new system, considering the European experiences and also the amount of effort currently being expended to modernize property-assessment files. In the New England states, where property assessment is split among as many as 40 or more cities and towns making up a single county, a multipurpose cadastre may be feasible only in the larger cities or for a county through intermunicipal agreements.

7.1.2 The Politics of Changing Local Systems

Organization of a multipurpose cadastre represents at least two fundamental changes in the general operating style of a local government. First, it is a recognition that the scope of responsibilities of the agencies of local government, taken together, has expanded to cover the total environment, and thus all of the land in its districts. Clerks who formerly kept records for use within their own offices become accountable for elements of the definitive land-record system of the district and may no longer be able to keep track of how and where their data are used. Second, it normally means the introduction of new technology into the local operations, with corresponding shifts in work loads, responsibilities, and skills required for individual jobs. The managers of the conversion process need to see themselves as agents of technological and institutional change.

The sharing of data that is assumed in the use of a multipurpose cadastre may lead to realignments of functions among the user agencies. If most of their functions already are automated, then the design of the new procedures for the multipurpose cadastre at least provides the occasion for review of the coding logic of the participating agencies.

Some of the agents of change will be operating outside of the county government. State legislatures will be involved in authorizing county agencies to carry out certain cadastral functions that may not be authorized at present. Typically, this is a matter of responding to initiatives of local officials who have drafted the bills that would provide them with the authority they seek.

The courts may be the ultimate agencies of institutional change in mandating the modernization of land-data systems, as they have been in so many other aspects of government in the United States. State constitutions typically require the updating of the assessed values that determine the distribution of real estate taxes at least every few years, and enforcement of this requirement often depends on the courts. Even an annual updating of property characteristics hardly seems enough to justify investing in a multipurpose cadastre. However, the most economical means of accomplishing it may be through a continuous updating of records as transactions occur, in a cadastral data system supported jointly with the other major users in the county.

7.2 POTENTIAL ROLES OF OTHER PARTICIPANTS OUTSIDE THE LOCAL GOVERNMENT

The potential roles of each of the five types of participants listed below should be studied carefully in the planning for development of a multipurpose cadastre. Few generalizations about them can be made here, because the sorting out of the functions of governmental services among these participants varies from one state to the next and among different sizes and types of communities within each state.

7.2.1 Intergovernmental Arrangements within the Region

Where the function of assessment of values of properties for taxation resides with the municipality, then this will likely be the most feasible location for the central management of the cadastral records system, as mentioned earlier. Many of the cities and towns in this situation find themselves with limited technical resources to develop and maintain the necessary systems of maps. Most of them must rely on the county deed recorder for ownership records. In these situations, the development of the multipurpose cadastre must be a joint effort of the county and the city or town.

Some regional agencies have assumed the function of maintaining land-parcel records for users in a multicounty district. Most regional planning agencies are eager to foster the development of multipurpose cadastres in their districts to help provide for their own information needs. The regional agency may be the best location of the technical assistance staff needed to coach the staffs of both the suppliers and the users of the cadastral records during the development process.

7.2.2 Private Data Bases in the Locality

Private enterprises that maintain files of land data will be important consumers of the data provided by a multipurpose cadastre but are unlikely to be suppliers, except when they operate as the equivalent of paid vendors under contract. If a private firm can justify an investment in obtaining and keeping a file of land data, then it is a corporate asset that affects the future profitability of the firm. Forest-products companies, the real estate industry, and utilities are examples. Because their data may be better and give them competitive advantages, they often become a vested interest against improvement of public data.

There are some private firms that operate as both consumers and suppliers of local government data, and for them some quid pro quo cooperative arrangements may be possible. A gas company may provide the detailed locations of its underground lines if it can then be given preferential access to the mapping system that records them along with the other underground facilities. Such public-private cooperative arrangements have been difficult to organize and even more difficult to maintain on a continuing basis. The members of the panel are aware of only one such arrangement that has survived, the Regional Mapping and Land Records (RMLR) Program described in Appendix A.4, which is still in its developmental stages. Elsewhere, ad hoc arrangements among utility companies to permit access to each other's maps by a third party at only the necessary locations seem to be as far as the companies will go in shared use of their land records.

The function of assuring the status of land ownership may continue to be performed by private attorneys and title insurance companies in the United States, regardless of the efficiencies of government-operated title-registration systems in many other countries. The report of the U.S. Department of Housing and Urban Development (HUD), published at the conclusion of its four-year study of methods to improve land-title recordation and registration (Office of Policy Development and Research, 1981), states on page V-26:

... conventional recording systems will no doubt remain the principal method of storing and recording land title documents, whatever the merits of registration systems, and, therefore, improvements are most likely to occur in conventional systems.

Nevertheless, the private-land-title industry is totally dependent on legal requirements for filing of records of title transfers with a public office as the means of generating most of the ownership records. The private companies also may use the indexing services of the public office to find the relevant records, to varying degrees, depending on whether an alternative private "title plant" is available to the firm in that locality. As a vehicle for improving these public indexing services, in the interests of controlling the costs of buying a home, the HUD report cited above recommends adoption by state governments of the model statute known as the Uniform Simplification of Land Transfers Act (USLTA), prepared by the National Conference of Commissioners on Uniform State Laws (1977). Among the several important features of this model legislation, the one listed first in the HUD report is parcel indexing, which has long been in use as the most efficient system for accessing documents in private title plants and in many jurisdictions where the Public Land Survey System exists. An up-to-date cadastral overlay such as described in this report is the key to maintaining such an index.

The USLTA was criticized shortly following its adoption in 1977 for its lack of simplified, standard procedures for land-title recording beyond the provision of a geographic index (Pedowitz, 1978). A broader model statute for a multipurpose cadastre is needed (Cook, 1982).

7.2.3   State Operating Agencies

State agencies rely on local governments to provide the basic maps and records for pursuing a variety of programs, examples being site selection for public facilities, approval of environmental protection permits, and review of planning for a state-aided program. States also are concerned with equalization of real-property assessments, so that state formula grants to local governments can be equitable when they are tied to local tax effort and need.

However, one of the unfortunate breakdowns in the system of public records on land in this country is that the state operating agencies do not then use and support the record systems of the local government for keeping track of their actions. Typically, the environmental protection permits remain indexed by date or case number on the state agency's file, with no routine reference in the local files. The wealth of data assembled in environmental impact statements lies fallow after the decision in the same manner. Expensive, large-scale topographic mapping for design of a highway is oriented only for use in that job and not as an update of parts of the map system for that locality. Lacking a procedure for keeping the records, there is seldom any requirement that the state agency file ''as-built'' plans showing where the new facility actually was constructed. Those who need this detailed information must rely on the plans showing where it was intended to be built, if they can be found.

State operating agencies are among the largest land owners. The quality and completeness of their records on the public land are important to the other land owners and users in the vicinity. A state-level land-records program should have authority to assure that records of state-owned land meet standards that support the needs of the locality and its cadastral records system and not just the needs of the operating agency.

7.2.4   State Coordinating Agencies

The Office of Land Information Systems or its equivalent, as recommended for each state in the report of the Committee on Geodesy (1980), should look on the cadastral records offices of the counties (or cities) as its primary constituency. Each local office will be organizing on its own schedule, so that a sudden rush of new programs is unlikely, especially at the outset. Ample time should be available for testing the approaches and proposed standards of the statewide program.

We recommend that a designated state agency participate in the development of each local cadastre, reviewing its plans, providing technical and financial assistance, and monitoring the adherence to its guidelines for local programs. The guidelines should be advanced to the status of standards after there has been adequate opportunity to test them and to verify their conformance to relevant standards that may exist at higher levels of government.

When state operating agencies generate information concerning the land, as described in the preceding section, much of the field measurement and mapping is performed by private firms under contract with the state or local government. This provides an opportunity for the designated state or local Land-Information Systems office to review the contract as a routine administrative step, with approval of that office required for any portions of a state contract that will generate data within the scope of the local cadastre.

We recommend that state governments require that the Office of Land Information Systems, or its equivalent, approve the relevant portions of any contracts involving production of field measurements, maps, or other land data that are within the scope of the local cadastre.

The private firms that provide telephone, electric, gas, and, in some areas, water or other communications services, as regulated public utilities have requirements for keeping track of their facilities that parallel those of the local public works department. Like the public offices, the managements of these regulated monopolies do not have sufficient incentive from just their own operations to join with others in building a shared land data base. On the other hand, they usually have been able to raise many more millions of dollars to invest in their separate mapping and record systems than have their counterparts who operate the county government facilities. The executives of the regulated utilities have been able to justify far heavier investments in land data to their public regulatory boards than have the county executives to their councils or to the electorate. One would hope this is due, in part, to the recognition of high professional standards in the management of the regulated utilities. However, one must also recognize that the appointees to state regulatory boards do not answer to the voters for the duplication they permit between separate utility companies in the same way that county officials must answer for duplication they permit among county departments. The marginal costs of duplication, or the marginal savings from a shared data system, revert to the public in either case, through their utility bills or their property-tax bills.

We recommend that the regulated public utilities be required to identify the field measurements, maps, or other land data for which budgetary approval is being requested and that, prior to this approval, the state boards that regulate them be required to submit this information to the designated state and local offices responsible for local Land-Information Systems for evaluation of the adequacy of the data that should be included in the multipurpose cadastre and for a determination of any duplication of effort.

7.2.5 Federal Programs

The original basis of title to most of the land in the United States was as federal territory, and most of this has been subdivided and transferred to others with reference to the Public Land Survey System (PLSS). The major exceptions are the lands of the original 13 colonies, which make up 18 of our present states. Thus, the General Land Office of the U.S. Government was predominant in establishing the cadastral records for most of Ohio (where the system originated) and for all of 29 states that lie farther to the west—this number also includes Florida but not Kentucky, Tennessee, Texas, or Hawaii. The localities in these 30 states have inherited the PLSS as the base for referencing original title to their land. In nearly two thirds of these states the federal government has "closed" its work on the PLSS and has turned over to the state governments the responsibility for whatever efforts are made to maintain the property-referencing system. Variations have evolved among these states in the procedures required for identical survey tasks, such as relocation of the center point of a PLSS section. In the 12 PLSS states where this transfer of responsibility has not yet been made, the Cadastral Survey Office of the U.S. Bureau of Land Management (formed in 1946) retains the authority for maintaining the network of survey control monuments at the PLSS section corners and quarter-section corners and the records of the disposition of the land. Positioning of these corners in Alaska, for example, is currently a major federal program.

A separate federal program, currently administered by the National Geodetic Survey (NGS), has put in place the first- and second-order geodetic control points in a national network that provides a base for the geodetic reference framework for local cadastres, as described in Chapter 2. Federally sponsored research has been instrumental in the technological developments for positioning of points in the geodetic reference framework (see Section 2.3) and for mapping (see Section 3.5). Demonstration projects to determine the feasibility of new technology in typical local situations also have been sponsored on occasion by federal agencies, for example, the applications tested recently in Colorado by the Bureau of Land Management (BLM) (Hendrix, 1981).

Federal land-survey procedures continue to predominate in rural areas of the 30 PLSS states, but the development of standards for subdivision of land into urban lots has been largely independent of them. "Subdivision control" procedures were originally invented in the nineteenth century as a means of assuring the adequate description of newly created lots for title records. They have been established in most of the states since 1945, with the additional objectives of better site planning and assurance that adequate infrastructure is provided by the developer. Standards for survey and description of the lots normally are mentioned. In Wisconsin, for example, a standard established in 1848 that measurements around the boundaries of the new land subdivisions or any part thereof must close with an error no greater than 1:3000 remains in force today. There may be a dual set of standards for performance and recording of surveys for rural versus urban land within the same county, because the county government normally has much more involvement in the latter through its administration of the subdivision control process. An example is the current situation of the cadastral records in Jefferson County, Colorado (see Appendix A.3).

The federal government remains an important land owner in the western third of the continental United States, essentially from the Rocky Mountains westward, and in certain other parts of the nation that have national forests, military installations, and other government facilities. The Cadastral Survey Office of the BLM has the ultimate responsibility for locating the boundaries of these federal lands, which can have a major impact on other land owners in the locality, both as abuttors and as users of monuments placed by the BLM for other cadastral surveys in the vicinity. The need of the BLM for more resources to carry out this boundary-marking function properly has been the primary force behind legislation introduced by Senator Domenici (1981).

Rapid aggregation of land-ownership data is required for national perspective on a number of other issues that are of concern to the federal government. The recent study of systems for monitoring foreign ownership of U.S. real estate provides one example (U.S. Department of Agriculture, 1979).

A new, national digital coordinate PLSS data base is recommended by the Committee on Integrated Land Data Mapping (1982). Should this now become an objective of the federal government, it would generate a strong federal interest in the proper relocation of PLSS corners in local land surveys and collection of the data needed in the new federal data base.

Meanwhile, the federal program that would have a greater impact than any of those described above on the realization of local multipurpose cadastres would be grants in aid tied to the use of recognized standards for each component of the new systems, as described in Section 7.5.5.

7.3 SUMMARY OF COSTS FOR PROTOTYPICAL COUNTIES

Typical costs for the components of a multipurpose cadastre are listed in Chapters 2, 3, and 4 and in the Appendixes, per square mile of land covered and per parcel for alternative levels of precision, depending on local needs. To summarize the costs for a county government requires knowledge of these overall dimensions of the program plus a host of other factors such as the adequacy of existing control points, base maps, and cadastral parcel maps. In this section, summaries of costs are listed for two hypothetical sizes of counties to describe a range of total program costs that will have at least some relevance for most counties in the United States.

The sequence of steps assumed in these prototypes is only one of many sequences possible in building the cadastral records systems. Individual counties may choose to invest first in organizing their files of land-parcel characteristics and postpone the new mapping to a later phase, if it is urgent to realize early benefits in current administrative and tax-assessment programs.

7.3.1 The Dimensions of Prototypical Counties

The average land area of one of the 3114 counties and county equivalents of the United States, if one excludes Alaska, is about 956 square miles. The typical county has been subdivided into the sections and quarter-sections of the PLSS, even though many of the monuments that located this basic spatial reference framework for cadastral parcels have been lost or obliterated. A substantial part of the nation does not fit this rule, that is, the territories that belonged to the original 13 states and did not become part of the public domain and the states of Texas and Hawaii. These "nonconforming" states comprise 20 percent of the nation's land area where 44 percent of its population resides. However, there is no alternative description of the overall pattern of land subdivision that can serve as a generalization for them. Both of the hypothetical counties therefore are assumed to be subdivided according to the PLSS. Counties in the other 20 states probably can support their uneven patterns of property boundaries with a lower overall density of survey control points than that described in the following paragraphs for the PLSS states and thus with a lower expenditure per square mile for survey control, except in urban areas.

The density of survey control points would be four per square mile throughout the nonfederal lands of the PLSS states, with the spacing at half-mile intervals recommended in Section 2.2.3. Establishing this density of control will be the major component of the cost of a countywide multipurpose cadastre, ranging from about half of the total costs for the prototype urban county to about two thirds for the prototype rural county. However, because these points are the legally established reference network for cadastral parcels, the cadastral overlay will fall into place much more quickly once they are located. In rural areas covered by the PLSS, much of the cost of the control surveys actually could be attributed to the cadastral surveying component. A more detailed explanation of the need to establish the locations of the quarter-corners of the PLSS sections as control points is presented in the report of the Committee on Integrated Land Data Mapping (1982). That report describes a similar program of monumentation for federal lands, which would be under the direction of the BLM.

For the mapping component of a countywide multipurpose cadastre the costs will depend on how much of the land must be covered with each of the customary scales of maps listed in Section 3.4, which in turn depends on whether the land is subdivided at urban, suburban, or rural densities. Table 7.1 recapitulates three of

TABLE 7.1   Densities of Development in Prototype Counties

Type of Area	Range of Lot Frontages in Densest Sector	Customary Base-Map Scale	Square-Miles Suggested for Prototype Counties	
			Urban	Rural
Urban	50' to 90'	1:1200	96	4
Suburban	100' to 180'	1:2400	260	12
Rural	200' and greater	1:4800	600	940
TOTAL COUNTY			956	956

TABLE 7.2   Suggested Factors for Cost Estimations

Factors	Estimates for Prototypical Counties	
	Urban County	Rural County
Total population	500,000	20,000
Population per parcel	2.5	1.33
Total number of parcels	200,000	15,000

the customary scales given in Table 3.2 and suggests how much of each of the 956-square-mile prototype counties might need to be mapped at each scale. The presence of even a relatively small sector of denser development may require shifting to a larger scale of map, even though as much as 90 percent of certain map sheets may be less subdivided and mappable at the smaller scale. Some counties will have significant amounts of land subdivided into either larger or smaller parcels than are covered by this range. Where many lot frontages are 40 feet or less, a scale of 1:600 is suggested, as indicated in Section 3.4 of this report. Also, in "resource areas" that essentially are unsubdivided, maps at the scales of 1:12,000 or even 1:24,000 may be appropriate. However, neither of these extremes is included in the two hypothetical prototypes.

Estimates of costs of the cadastral overlay component of a multipurpose cadastre are normally expressed per parcel and computed on the basis of the total number of parcels in the mapped area. For these hypothetical cost summaries, the prototype urban county is suggested as having 200,000 parcels, and the prototype rural county, 15,000 parcels. These figures are consistent with populations of about 500,000 and 20,000 for these two counties, respectively, as indicated by the ratios of population per parcel given in Table 7.2. The latter appear typical of similar counties described in the report on the HUD Land Title Systems Demonstration (U.S. Department of Housing and Urban Development, 1981a, 1981b). Although the suggested population of 20,000 for a prototypical rural county may seem small, it actually is the median population of the 3143 counties and county equivalents in the United States as of 1975.

7.3.2   Summary of Cost Estimates for Prototypes

Appendix A.1 describes a program of relocating and monumenting of quarter-corners of the PLSS sections in the Southeastern Wisconsin Region (which includes the Milwaukee metropolitan area) that provides a model to which many counties can relate. Typical costs for recovering and preparing a PLSS corner were about $200 in 1980 dollars, which included relocation, monumentation, establishing witness marks and related ties, and documentation. The subsequent costs of accurately determining the location and elevation of each corner as a control point averaged about $400, for a total cost of about $600 per corner in 1980 dollars for the geodetic reference framework. For one of the prototypical counties, with an area of 956 square miles and thus something close to 3800 section corners and quarter-corners to be located as control points, this would indicate a total cost of about $2.28 million (see Table 7.3).

The costs described above are for the use of traditional ground-traverse methods for control surveys, with electronic distance-measuring equipment. This technology is available throughout the nation and can be undertaken in small increments where budgets are limited. Should a county choose to invest more heavily in an accelerated program of control surveys, then the use of one of the newer technologies described in Section 2.3 of this report should bring down the cost per corner substantially for the larger program. For example, in Section 2.3.1 it is suggested that the cost of photogrammetric triangulation could be as low as one third that of traditional first-order ground traversing. The cost of $400 per corner for survey control work was for horizontal control by third-order, class I, traversing and vertical control by second-order, class II, leveling. If this could even be cut in half with photogrammetric triangulation, then the total cost of the program for a 956-square-mile county would be $760,000 less than the totals listed in Table 7.3.

The costs of base mapping at each of the three scales listed in Table 7.3 are drawn from the typical costs in the Southeastern Wisconsin Region (Appendix A.1)

TABLE 7.3   Cost Estimates for Prototypical Counties

Component of the Multipurpose Cadastre	Basis of Cost Estimates		Costs for Prototypical Counties (in thousands of dollars)	
	Unit Used	Unit Cost (1980 Dollars)	Urban	Rural
REFERENCE FRAMEWORK	Survey corner	$ 600	2,280	2,280
BASE MAPS				
Urban areas	Square mile	5,000	480	20
Suburban areas	Square mile	2,000	520	24
Rural areas	Square mile	800	480	752
TOTAL COUNTY			1,480	796
CADASTRAL OVERLAY	Parcel	10	2,000	150
TOTAL COSTS FOR HYPOTHETICAL COUNTY			5,760	3,226

and also from figures for the state of Missouri (see Section 3.2). They indicate that the total cost of a base-mapping program for the prototype urban county would be $1.48 million, and for the prototype rural county, about $796,000.

The average cost per parcel of $10 for the cadastral overlay also was drawn from Appendix A.1 and has the advantage of being an easy base figure for local adjustments. Costs per parcel would be much lower in rural areas that had not been subdivided significantly below the quarter-section level, because all of these would have been located precisely in the control survey phase of the local cadastre program.

7.4 PERSONNEL RESOURCES FOR MANAGEMENT AND STAFFING

The county or municipality that undertakes to build a multipurpose cadastre will need the professional talent within the office assigned this responsibility, first to oversee the design and initial organization of the system and then to manage its continuing maintenance and improvement. In addition to individuals knowledgeable in each of the application areas, the office will require people with expertise in (1) geodetic control surveying, (2) photogrammetry and cartography, (3) land surveying, and (4) data-base management systems.

The success of the program will depend on the high standards of professional competence set by the responsible local office and salaries that can attract and hold qualified professionals in these positions. The people hired for the managerial positions should have the breadth to grow with the program, in a field where technology is often changing. Without competent professional direction, much of the heavy expenditures in the development stages of the multipurpose cadastre will be wasted, and the accomplishment of the system actually may be set back by many years.

Many counties find that employment of professional consultants is the only means of obtaining the professional help needed at the early stages of a new program or in the period of rapid development when an unusual depth of talent is needed. However, unless the program managers already are well grounded in these fields, they should not try to select and manage contractors for a multipurpose cadastre program by themselves but should seek the advice and assistance of a supportive office, e.g., at the state level, if one exists.

We recommend that the office in each state that is assigned responsibility for county and municipal Land-Information Systems give a high priority to the setting of high standards of professional personnel in each local cadastre program and support them with recommendations for job descriptions and salary levels, with advice and assistance in designing their programs, and in selecting among available consultants, if needed.

A model state office performing these functions is the Land Records Management Program in the Department of Administration of the state of North Carolina.

The people with the required professional talents are available in most parts of the nation, given the state of the U.S. economy in mid-1982. However, if counties across the nation begin the development of multipurpose cadastres on a broad scale (which could occur if federal assistance is provided), then a scarcity of personnel may develop, especially at the managerial level.

We recommend that continuing support be given to the university-level programs that currently are preparing people for professional careers in control survey engineering, photogrammetry, cartography, and management of cadastral record systems.

7.5   FINANCING THE DEVELOPMENT COSTS

The several alternative approaches to financing the development of a multipurpose cadastre described in this section are based on the assumption that it is embedded in the administrative functions of the county government, or its equivalent, for reasons described in the opening paragraphs of this chapter. The multipurpose cadastre should not be undertaken until there is a commitment to its continuing maintenance and improvement after the development period, supported by the real estate tax base of the county or municipality. For financing the heavy front-end costs of the development period, at least four other possible sources also should be considered, as outlined in the following paragraphs.

7.5.1   Real Estate Taxes

Land owners are the most clearly identifiable constituency that is likely to benefit from a multipurpose cadastre. Further, the relative amount of benefit likely to accrue to any one owner relates somewhat to the valuation of his property. Also, from the point of view of the local government administration, one major justification of the cadastre is to operate the real estate tax system in a fair and efficient manner. A proportionate share of the cost of the cadastre could reasonably be charged to the real estate tax as an overhead cost of the tax system itself. The real estate tax, therefore, is an equitable source of funds for the long-term financing of the development of the cadastre, as well as for the continuing costs of maintenance and improvement.

We recommend that the real estate tax base of the locality be used as the source of funds for the net long-term financing of at least one quarter to one half of the front-end costs of developing the multipurpose cadastre, after the income from user charges has been accounted for.

The development of initial data acquisition and data entry costs may be allocated over a period of up to 20 years. The maintenance costs for a multipurpose cadastre are generally estimated at one twentieth of the above costs per year. Therefore, even after the multipurpose cadastre has been set in place, counties must budget for maintenance on a continuing basis.

There is a major risk in relying on local government taxation to support a development program that takes 10 or 20 years. Voter attitudes toward new public investments can swing back and forth in several cycles in that amount of time, and the work on the cadastre could be stopped by any one of the negative swings. The design of the program therefore should be as uncomplicated as possible, so that the prospective benefits of completing it will be obvious. Further, if the program can be covered by an intergovernmental agreement, with commitment of continuing support from state and federal as well as local governments, then it will be less vulnerable to changes in politics at any one of these levels.

7.5.2 User Charges

Users and other direct beneficiaries of improved land records ideally should pay in proportion to their benefit. Since the benefits are diffuse and the initial costs are very high, it is necessary for government to play the major role. All three levels of government are involved.

Unfortunately, the diffuse nature of benefits from the cadastre makes it difficult to assign specific amounts to the beneficiaries. In addition to being difficult to capture, these benefits accrue in the long run while the costs are high initially to develop the maps and capture the data. A proposed charge for use of the cadastral maps or records at least can be evaluated by the question: Does the proposed fee recover costs from those who actually use the system?

The users having probably the greatest stake in the quality of the cadastral records system are those engaged in transfers of property rights. A cadastre financing fee could be charged to these users in many areas as an add-on to the existing state or local transfer tax for recording of the transaction as a percentage of the value of the property transferred.

We recommend that a share of the cost of developing and maintaining the cadastre be covered through a real estate transfer tax collected at the point of recording of the transaction. States that already have such a tax should earmark at least a portion of the proceeds to support the state office of Land-Information Systems, including a program of grants for development of county-level cadastres, and increase the tax rate to provide sufficient funds for these purposes, if necessary.

If at least some of the states do not proceed to raise funds for multipurpose cadastres through a transfer tax, then a national tax for this purpose administered by the federal government should be considered, analogous to the incremental tax on gasoline that funds the interstate highway system. State administration of the transfer tax would be preferable, so that it could be coordinated with existing transfer taxes and with the administration of the real estate taxes, which vary from state to state. Further, a federal transfer tax would be viewed initially as a reinstatement of the federal tax stamps on real estate transactions, which has been discontinued, and would require an extensive public relations program. Nevertheless, a national transfer tax would be preferable to no funding at all for the development of the multipurpose cadastre.

The issuance of building permits in most states already is heavily loaded with procedures for enforcement of building and zoning codes and fees that support the local and state staffs to maintain and administer them. Nevertheless, those who invest in new construction or expansion of existing structures have a major stake in the accurate record of locations and conditions of their properties and of the districts where various types of building restrictions apply.

We recommend that the possibility of raising some of the funds needed for the cadastre through a surcharge on fees for building permits also be considered in each state.

Some funds for at least maintenance of the system also can be recovered through charges for hard copies of the records (maps or tabulations on paper or film) or for the connection of a terminal to the electronic files so that they might be read in a private office. Users of the existing deed record books in some counties are willing to pay as much as 50¢ or 75¢ per sheet for paper copies of the pages they need. This obviously covers more than the cost of paper and duplication, but income from such sources could only be a small fraction of the total budget for operating and maintaining the cadastre, without any consideration for development costs.

7.5.3   Joint Venture with Large Private Data Users

A multipurpose cadastre, kept constantly up to date, will duplicate a major share of the land data now maintained in separate systems by the title insurance industry and by the utilities. If these companies would make the commitment to pay even a fraction of the present costs of obtaining these data to the county (or municipality) for the information service of the multipurpose cadastre, this could provide for a major part of the public budget.

A major obstacle to such arrangements at present is the lack of a sufficient commitment on the part of the local government to convince the private data user that the proposed new system will be stable and permanent and that the data will indeed be available when he needs them. If it is not, the potential business losses would quickly exceed the amounts currently spent to maintain the separate, private system. It is unlikely that the major private users can be convinced to change this position and commit their support to a future multipurpose cadastre unless either (1) its development and operation are controlled by a policy board on which they have a major voice or (2) the federal government is committed to support the development and maintenance of the local cadastre that meets nationally recognized standards.

7.5.4   State Matching Funds

Participation of the state government in each county program to develop a multi-purpose cadastre with state matching funds would serve a number of state-level interests. It would increase the incentives for counties to undertake the programs, benefiting the state operating agencies that would use the results, as listed in Section 7.2.3. It would assure the counties that the designs of their cadastres are consistent with statewide standards and comparable with other jurisdictions, which should help keep down the prices charged by vendors who serve the statewide market. It would put the state coordinating agency in a strong position to ensure that its procedural requirements are being met, e.g., for review of all public investments in surveying and mapping by the local agency administering the cadastre.

We recommend that the government of each state provide funds to the agency designated to participate in the development of each local cadastre, sufficient for grants of state funds for at least 25 to 50 percent of front-end costs of developing the multipurpose cadastre. We recommend that the provision of these categorical matching funds for the development of each local cadastre continue at least until it has become an established part of the local government, such that the state agency is confident of its permanence.

7.5.5 Federal Matching Grants

A commitment of the federal government to support the development of a multipurpose cadastre to serve each county (or municipality, where appropriate) would lead to a new era for local land-data systems in the United States, providing a unity and direction that have been lacking to date. Without such a program, the development of cadastres in this country will continue to be piecemeal, sporadic, and slow, as it has been to date, even if the other standards and procedures recommended in this report are followed. Examples of all the components of a multipurpose cadastre being operated to high standards somewhere in the United States have been identified. However, we have been unable to identify any county or municipality that has had the resources to put all of them together and maintain them in a coordinated system.

We recommend that an agency of the federal government be designated to design and develop a program of federal financial assistance to counties (and certain municipalities) for the development of multipurpose cadastres along the lines described in Section 4.3.2 of the Committee on Geodesy (1980) report, with federal matching funds provided at a level of about 40 percent, contingent upon the minimum additional contribution of about 20 or 30 percent by the state government, with its active participation. The design and development work should include the drafting of proposed standards and procedures covering at least the scope of those presented in this report and testing of them in a series of demonstration projects within selected counties.

For a rough estimate of the order of magnitude of cost to the federal government for a program of matching grants, one might make the rough assumption that the cost of providing a complete cadastral system for the average county in the United States would be about $3.5 million if it proceeded individually, using the hypothetical figures developed in Section 7.3. However, the economies of high technology that

TABLE 7.4   Percentages of Total Estimated Costs to Be Borne by a Federal Grant Program

Factor	Estimated Percentage	Rationale
Exclusion of federal land	80	22 percent of the continental United States is owned by the federal government (area of Alaska not included in original cost estimate)
Use of existing control points and maps	75	25 percent of the cost might be saved by use of existing ground-control, base maps, and cadastral maps
Spread over 20 years	5	
Federal matching funds ratio	40	Requires a major share (60 percent) to be committed by state and local governments
COMBINED	1.2	Product of the four estimated percentages

should be possible in a nationally coordinated program should save at least $1 million of this amount. If cadastres for all the 3114 counties and county equivalents needing to develop cadastres (outside of Alaska) were developed with survey control points at one-half-mile intervals, the national total for coast-to-coast cadastres could be in the vicinity of $7.8 billion, in 1980 dollars. However, because the 24 percent of the area of the continental United States that lies outside the PLSS could make do with a lower density of control points, $7.5 billion would seem to be a reasonable estimate for total program costs, excluding Alaska. This compares reasonably with the estimate of $3.35 billion for a national program derived from the study by the U.S. Department of Agriculture (1979) and reported on in the Committee on Geodesy (1980) report, which did not include investment in the geodetic reference framework and was expressed in 1979 dollars.

A rough estimate of the annual contribution from the federal grants to counties and municipalities to support the program at the levels recommended in this report would be about 1.2 percent of the total estimated cost, or $90 million per year for 20 years, based on the factors in Table 7.4.

It should be clear from the above that the rough estimate of $90 million per year for 20 years suggested as appropriate for a program of federal matching grants for the multipurpose cadastre is not the result of research but rather of some rough approximations that seem reasonable. If there is general agreement on the parameters of a multipurpose cadastre described herein, then more accurate estimates could be developed fairly rapidly from data available from the U.S. Census of Governments. In the meantime, this rough estimate at least provides the starting point for the discussion of the conclusions presented in Chapter 8.

8
Recommended Activities at the National Level

Three fundamental components of a multipurpose cadastre have been identified in this report: (1) a geodetic reference framework, (2) a base map, and (3) a cadastral overlay. Only where these technical components are adequately provided can the development of the cadastre proceed on a sound basis and eventually support permanent linkage mechanisms among real-property title, fiscal, and administrative records. Moreover, only where these technical components are adequately provided can the multipurpose cadastre eventually be expanded to a multipurpose land-data system incorporating natural resource base and land-related socioeconomic data.

The outline of required standards and procedures presented is intended primarily for those who would support the organization, design, or administration of a multipurpose cadastre at the county level. This concluding chapter offers suggestions for steps that might currently be taken by federal agencies and national associations that support these recommendations.

There are many federal programs that could use a county-level cadastre in support of their operations, some of which were listed in Section 7.2.5. The cadastre would provide the vehicle for recognizing and preserving a high quality of cadastral surveys locating the boundaries of federal ownership and interests in land. A wealth of land data would be more readily available for site evaluations for energy facilities, for federal installations, for historic preservation, for management of agricultural programs, for development of natural resources, and for control of pollution. A common vehicle would be available for permanent recording of the decisions reached in these programs. National accounts could be more readily compiled for evaluations of national assets and who controls them.

Other federal programs will benefit from the cadastre not so much by using it as by depending on it to support their objectives directly. Examples are the programs of the Department of Housing and Urban Development (HUD) established by the Real Estate Settlement Procedures Act (RESPA), seeking lower costs for buying homes, and the many federal programs aimed at developing and employing the professional human resources of the nation.

8.1   CLEAR STATEMENTS OF OBJECTIVES NEEDED

Any federal initiatives in this field will be in an intergovernmental context. One of the first items on any federal agenda for a multipurpose cadastre should be to resolve a clear statement of the objectives of federal initiative, whether by one agency or an interagency consortium. Land-Information Systems can mean so many different things to different people that confusion of objectives is one of the greatest risks to success of a federal effort.

The list following this paragraph offers a few suggestions of the general areas into which the long-range objectives of a federal initiative might fall. The more immediate, short-range objectives are too varied and too specific to individual agencies and programs to attempt any listing here.

• Promote sharing of technology and of data through use of common standards for definitions of terms and for data quality.

• Encourage volume production of software and equipment, to realize lower costs.

• Encourage and support the establishment of centers of excellence in land-information science.

• Create opportunities for useful employment of young people with professional education.

• Speed up the delivery of benefits of the multipurpose cadastre in each locality, as detailed in Table 1.1, for example:

Better access for individual land owners and citizens to land records that may affect their personal interests.

Better informed public decisions through access to shared records of all public actions affecting specific land parcels.

More effective land-use planning and protection of scarce land-based resources through accurate records of land qualities and existing restrictions.

More fair and equal taxation of real estate through total accounting of real property.

Clearing up of confusions or inconsistencies in present records relating to adjacent land parcels.

More effective management of public lands.

Lower costs for public utilities through sharing of basic geographic data.

8.2 DRAFTING AND PROMOTION OF STANDARDS

Federal agencies are in a difficult position to proceed with the drafting and promotion of standards for a multipurpose cadastre without being accused of seeking to take over control of the data systems from state and local governments. However, by working with the designated representatives of the local governments through their national associations as participants from the very beginnings of a federal initiative, the initiative should be more informative and also more effective in gaining acceptance of the results by the county and municipal governments.

We urge the National Association of Counties (NACo), through its appropriate constituent organizations and staff, to organize a review of the findings and recommendations of this report that involves representatives of local user agencies and identify the areas in which more specific standards and procedures are most needed to make the approach described here operational. We urge that the federal offices with the technical skills required for defining standards for geodetic surveying, base mapping, cadastral mapping, and land-attribute data be invited by NACo to contribute to these processes at the appropriate points.

A number of other national associations also might play important roles in the research, drafting, or publication of recommended procedures and standards, depending on how NACo chooses to define its own role in this effort. The prime candidates are the other national associations and agencies that will be involved in the development, operation, and use of multipurpose cadastres, including those affiliated with the Institute for Modernization of Land Data Systems (MOLDS).

8.3 RECOGNITION OF STANDARDS BY FEDERAL AGENCIES

The Committee on Geodesy (1980) report recommended (in Section 4.3.2) that any federal agencies that produce or fund components of a multipurpose cadastre (such as right-of-way surveys or large-scale maps) should be required to adhere to a federal plan that establishes the "format" for these components or, until such a plan is adopted, to the individual state plan, if any. Given the materials presented in Chapters 2 through 5, the word "format" should be changed to read "procedures and standards" and should include the more detailed recommendations that might result from further initiatives by organizations such as NACo, as recommended in the preceding section.

There remains an urgent need for designation of a single lead agency in the federal government in the field of surveying and mapping to provide a structure for the formal recognition of procedures and standards for a multipurpose cadastre, as described above, and to oversee compliance with them by the federal establishment. The need for designation of such an agency was stated in the report to the Office of Management and Budget by the Federal Mapping Task Force (1973), was endorsed in the concluding chapter of the Committee on Geodesy (1980) report, and was reiterated in the Committee on Geodesy (1981) report.

8.4   ORGANIZING A PROGRAM OF FEDERAL ASSISTANCE

Standards and procedures of the scope recommended in this report will not by themselves assure that cadastres will be organized on any broad scale. If the development of a multipurpose cadastre is left to wait for local leadership, the results will be slow to be realized, disjointed, and of uneven quality. There will be a risk of general rejection of this approach across the nation if the first few localities to attempt it are not successful.

The risks of such failures can be minimized with an adequate federal commitment to follow through with support for the cadastres and to maintain high standards of quality for the programs that are assisted. The stability that a federal assistance program would lend to development of local cadastres will foster many of the other long-range commitments that are important to its success, such as participation by private utility companies and the attraction of talented young people into the professional fields where they will be needed. The federal-aid highway program provides an example of the potential effect of such a program in generating high standards of professional work and productivity in the responsible offices of state governments throughout the nation.

The dimensions of a recommended program of federal assistance (see Section 7.5.5) with an estimated annual budget of $90 million, which represents 40 percent of the annual costs, are vaguely defined in this report. There is much that can be done at present to define more specifically how the program would work in states and counties, which would be bearing 60 percent of the costs (see Table 7.4), and to build a plan and budget for a federal program with enough specifics to be considered by the Congress.

Appendix A
Laying the Technical
Foundation for a
Multipurpose Cadastre:
Referrals and Case Studies

The experiences of counties and municipalities that have begun to build their multipurpose cadastres is instructive, especially for understanding the scenario of decisions and investments that can be successful. This report has concentrated on the objectives of the program: the scope and quality of operations and products eventually to be accomplished. However, a number of city, county, and regional agencies have succeeded in establishing major components of their multipurpose cadastres, and reports on most of them are available in the literature. The programs of four of them are summarized briefly in this Appendix. Those who may wish additional information on these, or other local programs worthy of attention, may find the list of publications and contacts below to be useful. This is not intended as an exhaustive list of such programs; it presents those projects that were available to the authors during the study.

Each county or municipality that undertakes a multipurpose cadastre begins with a unique set of needs and resources. Some of the key determinants of its program will depend on the answers to the following questions:

- What is the status of the existing system?
- What are the objectives of the users of the system?
- What standards and procedures are required to meet the desired objectives or uses?
- What are the costs and benefits?

Programs Described in the Case Studies in this Appendix

A.1 The Southeastern Wisconsin Region

For Further Reference: K. W. Bauer, Integrated large-scale mapping and control survey program completed by Racine County, Wisconsin, *Surveying and Mapping 36*(4) (December 1976).

Contact: Kurt W. Bauer, Executive Director
Southeastern Wisconsin Regional Planning Commission
916 North East Avenue
Waukesha, Wisconsin 53187
(414) 547-6721

A.2 DuPage County, Illinois

For Further Reference: J. G. Donahue and W. J. Faedtke, DuPage County, Illinois, Remonumentation and integrated computer mapping program, *Surveying and Mapping 42*(2), 113-124 (1982).

Contact: William J. Faedtke
DuPage County Center
421 N. County Farm Road
Wheaton, Illinois 60187
(312) 682-7000

A.3 Jefferson County, Colorado

For Further Reference: B. C. Swenson, Graphic Systems Director, Jefferson County Mapping Division, 1010 Tenth Street, Golden, Colorado 80401, personal communication (February 1982).

Contact: Billie C. Swenson, Director
Jefferson County Mapping Office
1010 Tenth Street
Golden, Colorado 80401
(303) 277-8308

A.4 Philadelphia Area

For Further Reference: Delaware Valley Regional Planning Commission, *The Regional Mapping and Land Records Program—A Summary Report*, Philadelphia, Pennsylvania (July 1980).

Contact: John M. Hadalski, Jr.
Chairman, RMLR Steering Committee
Office of the Managing Director

City of Philadelphia
1620 Municipal Services Building
Philadelphia, Pennsylvania 19107
(215) 686-7114

or   Roger Smith
Delaware Valley Regional Planning Commission
1819 J. F. Kennedy Boulevard
Philadelphia, Pennsylvania 19103
(215) 567-3000

Other Programs

1.   Forsyth County, North Carolina

Reference: E. Ayers, Developing Necessary Political Support for a Modern Land Records System, Proceedings of the 20th Annual Conference of the Urban and Regional Information Systems Association, August 1982, published by URISA, 2033 M Street, N.W., Washington, D.C. 20036.

Contact: John W. Jones
Data Processing
Forsyth County
Winston-Salem, North Carolina 27102
(919) 727-2597 or 727-2167

2.   Lane County, Oregon

Reference: Lane County Regional Information System, *Regional Information System Long Range Plan*, 1981-1986. Available from Lane Council of Governments, 125 Eighth Avenue East, Eugene, Oregon 97401 (January 1982).

J. R. Carlson, ADLIB: A Multi-Function Site Address Library, Proceedings of the 20th Annual Conference of the Urban and Regional Information Systems Association, August 1982, published by URISA, 2033 M Street, N.W., Washington, D.C. 20036.

Contact: James R. Carlson
Lane Council of Governments
125 Eighth Avenue East
Eugene, Oregon 97401
(503) 687-4283

3.   North Carolina Land Records Program

Reference: North Carolina Department of Administration, *Keys to the Mod-*

ernization of County Land Records, Land Records Management Program, Raleigh, North Carolina (1981).

Contact: Donald P. Holloway, Director
North Carolina Land Records Management Program
Department of Administration
116 West Jones Street
Raleigh, North Carolina 27611
(919) 733-2566

4.   Wyandotte County, Kansas

Reference: Wyandotte County Base Mapping Program, *Development of a Multipurpose Cadastre in Wyandotte County, Kansas*, Wyandotte County Government, Kansas City, Kansas (1982).

Contact: D. Edward Crane, Project Director
Wyandotte County Base Mapping Program
County Court House
Kansas City, Kansas 66101
(913) 573-2941

Appendix A.1
The Southeastern Wisconsin Region

Recognizing the importance of good large-scale maps to sound community development and redevelopment, the Southeastern Wisconsin Regional Planning Commission (referred to below as "the Commission") has, for two decades, encouraged the preparation of large-scale topographic and cadastral maps within its 2689-square-mile region. These maps are based on a unique system of survey control that combines the best features of the U.S. Public Land Survey System and State Plane Coordinate Systems. The large-scale maps and attendant survey control system provide, in a highly cost-effective manner, the technical foundation for the eventual creation of a multipurpose cadastre within the region.

THE SURVEY CONTROL FRAMEWORK

The Commission is committed to the concept that any accurate mapping project requires the establishment of a basic system of survey control. This control consists of a framework of points whose interrelationships and whose horizontal and vertical positions on the surface of the Earth have been accurately established by field surveys and to which the map details are adjusted and against which they can be checked.

At present, new large-scale topographic mapping in most urban areas is usually based on third- or lower-order control nets having, at best, temporarily monumented stations. These control nets are usually largely unrecoverable and, as a practical matter, unusable by local engineers and surveyors. These control nets are generally tied to the national geodetic datum, and the finished maps are compiled on a state plane coordinate grid. Property boundary-line maps are, on the other hand, most often mere compilations of paper records, no real framework of control or map projection being utilized in their construction at all. In such situations, the accurate correlation of cadastral maps with topographic maps and even with other cadastral maps is manifestly impossible.

A comprehensive system of horizontal control based on the U.S. Public Land Survey System, as well as on the national geodetic datum, has, therefore, been proposed and utilized by the Commission as a basis for the compilation of large-scale maps that are adequate for planning and engineering purposes. The establishment of such a control system requires the relocation and monumentation of all section and quarter-section corners within the area to be mapped and the utilization of these corners as stations in a third-order, class I,* traverse net tied to the national geodetic datum. Although this order of accuracy is not required for the map production, it is required if the control net is to have permanent utility for all subsequent local survey work.

The control traverse net establishes the exact lengths and bearings of all U.S. Public Land Survey quarter-section lines, as well as the geographic positions, in the form of state plane coordinates, of the Public Land Survey corners themselves throughout the area to be mapped. The elevations of the monuments marking the U.S. Public Land Survey corners are also determined by second-order, class II, level circuits.†

Six important advantages of this system of survey control developed by the Commission are stated in Section 2.3.2.

TECHNICAL PROCEDURES AND REQUIREMENTS

All the control survey work and attendant mapping have been carried out in accordance with a standard set of specifications provided by the Commission. These specifications call for the preparation of photogrammetrically compiled topographic maps that meet National Map Accuracy Standards at scales of 1:1200 or 1:2400, with a vertical contour interval of 2 ft, the maps being based on the herein described survey control system. Through the cooperative efforts of the Commission and certain county and local units of government, this survey control and mapping system to date has been extended into 1033 square miles, or over 38 percent of the total area of the region. A total of 5678 U.S. Public Land Survey corners have been relocated, monumented, and coordinated, representing over 48 percent of such corners in the region (see Figure A.1).

---

*Position closure after azimuth adjustment not to exceed 1 part in 10,000; azimuth closure not to exceed 3 sec of arc per station; field procedures, computations, and adjustments to follow National Geodetic Survey methods.

†Maximum error of closure in feet, 0.02/level circuit length in miles; field procedures, computations, and adjustments to follow National Geodetic Survey methods.

LARGE-SCALE TOPOGRAPHIC
MAPPING AND RELOCATION,
MONUMENTATION, AND
COORDINATION OF U. S. PUBLIC
LAND SURVEY CORNERS: 1980

LEGEND

☐ LARGE-SCALE TOPOGRAPHIC MAPPING
   COMPLETED OR UNDER PREPARATION

• U.S. PUBLIC LAND SURVEY CORNERS
   WHICH HAVE BEEN OR ARE BEING
   RELOCATED, MONUMENTED, AND
   COORDINATED

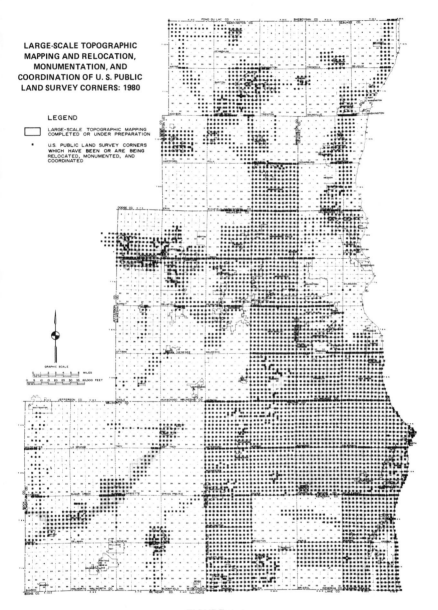

FIGURE A.1

The specifications governing the work require that the relocated Public Land Survey corners be marked by reinforced concrete monuments, having engraved bronze caps imbedded in the tops (see Figures A.2 and A.3). The bronze caps are stamped with the corner notation—quarter-section, town, and range.

The monuments placed are referenced by ties to at least three witness marks. The specifications require that the survey engineer provide a dossier on each control station established in order to permit its ready recovery and use. The dossier sheets are prepared on 8½-inch × 11-inch base material and provide for each station a sketch showing the monument erected in relation to the salient features of the immediate vicinity, all witness monuments together with their ties, the state plane coordinates of the corner, its Public Land Survey description, the elevation of the monument, and of appurtenant reference benchmarks referred to National Geodetic Vertical Datum of 1929 (see Figure A.4). These dossier sheets are recorded with the County Surveyor as well as with the Commission and are thereby readily available to all land surveyors and engineers operating in the area mapped.

The specifications require the control survey data to be summarized by means of a control survey summary diagram showing the exact grid and ground lengths and grid bearings of the exterior boundaries of each quarter-section; the area of each quarter-section; all monuments erected; the number of degrees, minutes, and seconds in the interior angles of each quarter-section; the state plane coordinates of all quarter-section corners together with their Public Land Survey System identification; the benchmark elevations of all monuments set; and the basic National Geodetic Survey control stations utilized to tie the Public Land Survey corners to the horizontal geodetic control datum, together with the coordinates of these stations. The angle between geodetic and grid bearing is noted, as is the combination sea-level scale-reduction factor (see Figure A.5).

All the work necessary to execute the control surveys and provide the finished topographic maps described below has been done in southeastern Wisconsin on a negotiated contract basis with a photogrammetric and control survey engineer. In this regard it was considered essential to retain a photogrammetric and control survey engineer familiar with higher-order field methods and procedures and with the attendant geodetic survey computations and adjustments and whose crews were properly equipped with state-of-the-art survey instruments. Electronic distance-measuring equipment was employed in the work, as well as optically reading theodolites and appurtenant traverse equipment, automatic levels, and precision level rods. Indeed, the control survey system used is made economically feasible only through the application of these relatively recently developed instruments, particularly the electronic distance-measuring devices.

Although the specifications governing the work make the photogrammetric engineer responsible for overall supervision and control of the mapping work, as well as for the quality of the finished maps, they require that the actual relocation of the Public Land Survey corners be done by a local land surveyor employed as a sub-

DETAIL OF MONUMENT AND MONUMENT INSTALLATION
FOR SURVEY CONTROL STATIONS

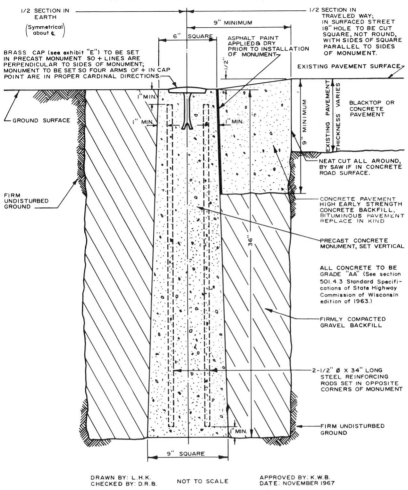

1/2 SECTION IN EARTH
(Symmetrical about ₵)

1/2 SECTION IN TRAVELED WAY; IN SURFACED STREET 18" HOLE TO BE CUT SQUARE, NOT ROUND, WITH SIDES OF SQUARE PARALLEL TO SIDES OF MONUMENT.

9" MINIMUM

6" SQUARE

ASPHALT PAINT APPLIED & DRY PRIOR TO INSTALLATION OF MONUMENT

BRASS CAP (see exhibit "E") TO BE SET IN PRECAST MONUMENT SO + LINES ARE PERPENDICULAR TO SIDES OF MONUMENT; MONUMENT TO BE SET SO FOUR ARMS OF + IN CAP POINT ARE IN PROPER CARDINAL DIRECTIONS

EXISTING PAVEMENT SURFACE

GROUND SURFACE

1" MIN

1" MIN

1" MIN

BLACKTOP OR CONCRETE PAVEMENT

EXISTING PAVEMENT THICKNESS VARIES

9" MINIMUM

NEAT CUT ALL AROUND, BY SAW IF IN CONCRETE ROAD SURFACE.

FIRM UNDISTURBED GROUND

CONCRETE PAVEMENT HIGH EARLY STRENGTH CONCRETE BACKFILL, BITUMINOUS PAVEMENT REPLACE IN KIND

PRECAST CONCRETE MONUMENT, SET VERTICAL

ALL CONCRETE TO BE GRADE "AA" (See section 501.4.3 Standard Specifications of State Highway Commission of Wisconsin edition of 1963.)

36"

FIRMLY COMPACTED GRAVEL BACKFILL

2-1/2" Ø X 34" LONG STEEL REINFORCING RODS SET IN OPPOSITE CORNERS OF MONUMENT

FIRM UNDISTURBED GROUND

1" MIN.

9" SQUARE

DRAWN BY: L.H.K.
CHECKED BY: D.R.B.

NOT TO SCALE

APPROVED BY: K.W.B.
DATE: NOVEMBER 1967

Source: SEWRPC.

FIGURE A.2

DETAIL OF TYPICAL ALTERNATIVE CONTROL SURVEY MONUMENT INSTALLATION IN
SURFACED TRAVELED WAY OF STREETS AND HIGHWAYS

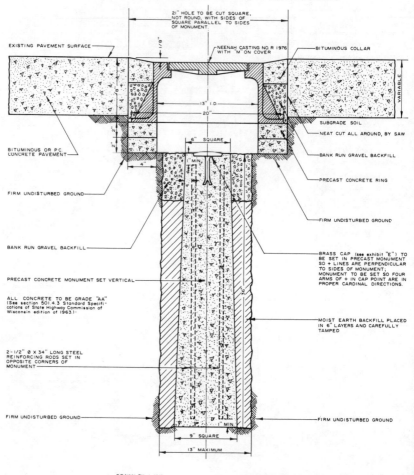

FIGURE A.3

RECORD OF U.S. PUBLIC LAND SURVEY CONTROL STATION

QUARTER SECTION CORNER 30|30 / 30|30 T_2_ N, R_23_ E, _KENOSHA_____COUNTY, WISCONSIN

GEODETIC SURVEY BY: _ALSTER-AYRES & ASSOCIATES, INC._

STATE PLANE COORDINATES OF: _CENTER OF SECTION_
NORTH _227,226.44_
EAST _2,585,319.67_

ELEVATION OF STATION: _622.17'_____ THETA ANGLE: _+ 01-29-38_

HORIZONTAL DATUM: WISCONSIN STATE PLANE COORDINATE SYSTEM, SOUTH ZONE

VERTICAL DATUM: MEAN SEA LEVEL, 1929 ADJUSTMENT

HORIZONTAL & VERTICAL CONTROL ACCURACY: SECOND ORDER

LOCATION SKETCH:

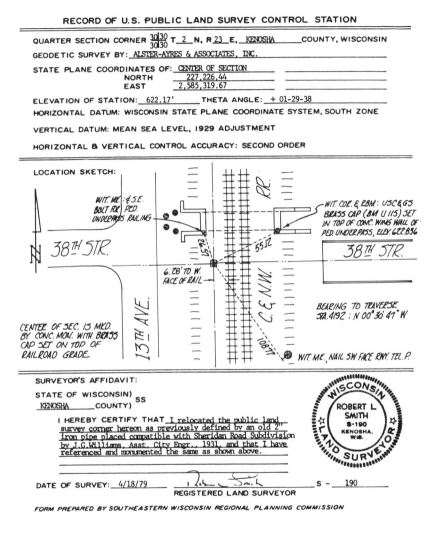

SURVEYOR'S AFFIDAVIT:

STATE OF WISCONSIN) SS
_KENOSHA_____COUNTY)

I HEREBY CERTIFY THAT _I relocated the public land survey corner hereon as previously defined by an old 2" iron pipe placed compatible with Sheridan Road Subdivision by J.G.Williams, Asst. City Engr., 1931, and that I have referenced and monumented the same as shown above._

DATE OF SURVEY: _4/18/79_   _Robert Smith_   S - _190_
REGISTERED LAND SURVEYOR

FORM PREPARED BY SOUTHEASTERN WISCONSIN REGIONAL PLANNING COMMISSION

FIGURE A.4

140

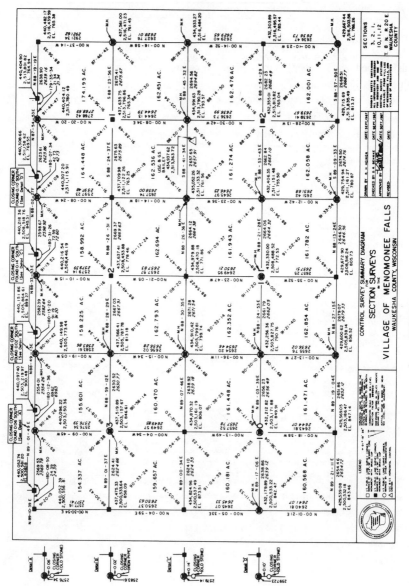

FIGURE A.5

contractor by the photogrammetric engineer. The specifications thereby recognize that this portion of the work requires expert knowledge of local survey custom and boundary and title law, as well as the assembly and careful analysis of all authoritative survey information—such as title documents, subdivision plats, survey records, and, of cardinal importance, existing monumentation and occupation—in order to arrive at the best possible determination of the location of the land-survey corners. In the areas mapped, the land-survey portion of the control survey work requires a very high degree of professional competence as almost all of the Public Land Survey corners fall under the federal definition of either obliterated or lost corners. The importance of this phase of the work and its impact on real property boundaries throughout the community can hardly be overemphasized.

BASE MAPS—SUGGESTED USE OF LARGE-SCALE TOPOGRAPHIC MAPS

The specifications provide for the completion of finished topographic maps that can serve as the base maps for the preparation of a multipurpose cadastre by accurately recording the basic geography of the area mapped. In addition to showing the usual contour information, spot elevations, planimetric and hydrographic detail, and coordinate grid ticks, the maps show, in their correct position and orientation, all U.S. Public Land Survey quarter-section lines and corners established in the control surveys (see Figure A.6). The specifications require that the maps be prepared to National Map Accuracy Standards. Thus, all state plane coordinate grid lines and tick marks and all horizontal survey control stations must be plotted to within 1/100 inch of the true position as expressed by the adjusted coordinates for the control survey stations, and 90 percent of all well-defined planimetric features must be plotted to within 1/30 inch of their true positions, and no such features may be off by more than 1/20 inch. Ninety percent of the elevations indicated by the solid-line contours must be within one-half contour interval of the true elevation, and no such elevation may be off by more than one contour interval. A combination sea-level and scale-reduction factor, and the angle between geodetic and grid bearing, are noted on each map sheet, as is the equation between any local datum and mean sea level.

Importantly, all finished maps are field checked by the Commission. This check involves the field inspection of all control survey monumentation and the running of traverse and level lines to verify the accuracy of the basic control surveys, as well as of the map details.

CADASTRAL OVERLAY

Actual property boundary-line maps, complementing the topographic maps, are compiled by the respective local units of government, utilizing resident engineering and

142

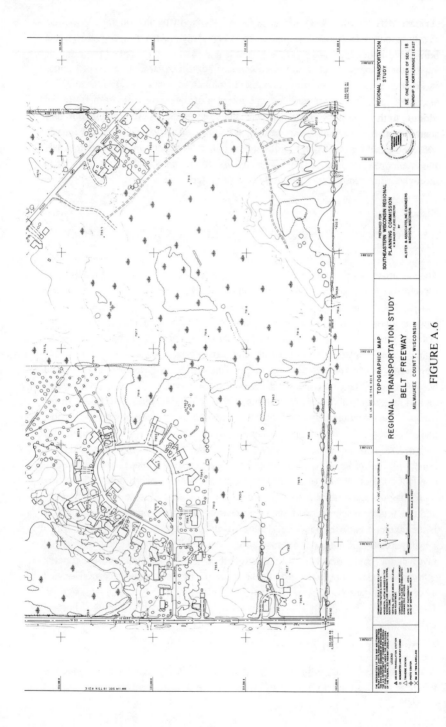

FIGURE A.6

planning staffs or consultants. Property boundary-line maps are compiled at a scale matching that of the topographic maps, each map sheet covering a U.S. Public Land Survey section or quarter-section.

As the topographic maps are being compiled, the specifications require that the photogrammetric engineer provide cadastral base sheets. These sheets consist of reproducible duplicates of the partially completed topographic maps showing, in addition to the state plane coordinate grid, the U.S. Public Land Survey section and quarter-section lines and corners in their correct position and orientation, together with their exact ground lengths and grid bearings, and such salient planimetric detail and hydrographic features as may be helpful in the subsequent plotting of real-property boundary lines, including roadway pavements, railway tracks, electric-power transmission lines, principal structures, fences, wetlands, lakes, streams, and drainage ditches.

Utilizing recorded subdivision plats, certified survey maps, and legal descriptions, all real-property boundary lines, including street right-of-way lines and utility easement lines, are constructed on the base sheets working within the framework of control provided by the ground lengths and grid bearings of the U.S. Public Land Survey quarter-section lines. The property boundary lines are constructed in a manner that parallels the location of these lines on the surface of the Earth following land-surveying practice in the state of Wisconsin. The specifications require that all real-property boundary lines be plotted within 1/30 inch of their true position based on analysis of all authoritative information available. Dimensions are shown for all platted areas as shown on the recorded subdivision plats. Wisconsin statutes have long required that such plats be prepared to an accuracy of 1 part in 3000 as compared with the accuracy of 1 part in 10,000 required by the specifications for the basic survey control network. Any overlaps or gaps between adjoining property boundary lines, as indicated by the constructions and plotting of those lines, are noted on the cadastral maps. Finally, a cadastral parcel number is added, thus providing the basis for the development of the linkage mechanism necessary for the creation of a multipurpose cadastre.

The property boundary-line maps thus show the ground length and grid bearing of all quarter-section lines; the state plane coordinates of all quarter-section corners; the monuments marking these corners; the recorded dimensions of all street lines, alley lines, and boundaries of public property; recorded street widths; platted lot dimensions; and a parcel-identification number. In unplatted areas real-property boundaries are shown by scale alone. Roadway pavements, railway tracks, electric-power transmission lines, principal structures, fences, wetlands, lakes, streams, and drainage ditches are also shown (see Figure A.7). As previously noted, these boundary-line maps can be readily and accurately updated and extended as new land subdivision plats and certified survey maps, utilizing the survey control, are made and recorded (see Figure A.8).

The cadastral overlays can be readily converted to digital form using an interactive

144

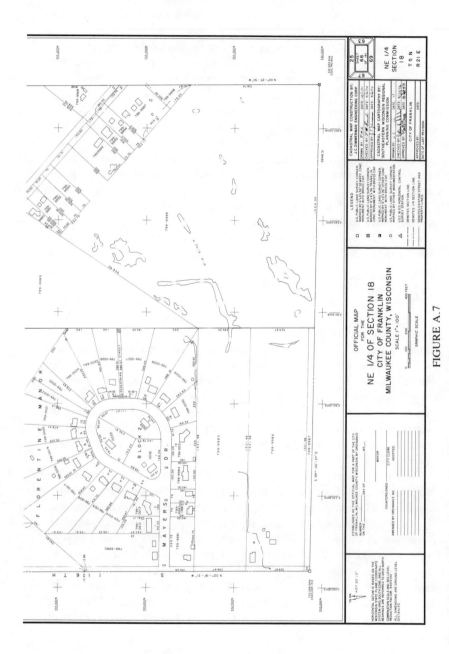

FIGURE A.7

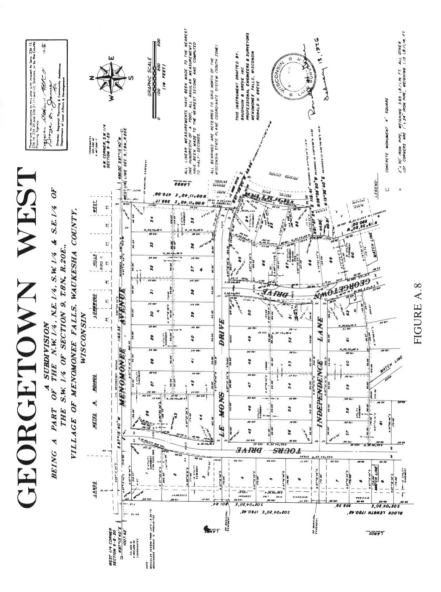

GEORGETOWN WEST

A SUBDIVISION
BEING A PART OF THE N.W.1/4, N.E.1/4, S.W.1/4 & S.E.1/4 OF
THE S.W. 1/4 OF SECTION 9, T.8N, R.20E.,
VILLAGE OF MENOMONEE FALLS, WAUKESHA COUNTY,
WISCONSIN

FIGURE A.8

graphic digitization and display system. This latter step toward the creation of an automated multipurpose cadastre has been accomplished only on a pilot basis within southeastern Wisconsin. A copy of a machine-produced cadastral map at a scale twice the base-map scale is shown in Figure A.9. The parcel-identification number serves as an index linking the parcel to title, tax, and public-land use regulatory information on file in various departments of county government.

Compilation of the property boundary maps in the manner described permits their reduction on a 10-to-1 ratio for the compilation of an accurate wall map at a final scale of 1:12,000 by mosaic process and at a 2-to-1 ratio for compilation of base maps for land subdivision planning and systems-engineering purposes. Contour information is, of course, readily and accurately transferable from the topographic maps by a simple overlay process.

In Wisconsin the mapping procedure is carried one step further. Section 62.23(6) of the Wisconsin statutes provides that the Common Council of any city* may establish an official map for the precise designation of right-of-way lines and site boundaries of streets and public properties. Such a map has all the force of law and is deemed to be final and conclusive as to the location and width of both existing and proposed streets, highways, and parkways and as to the location and extent of existing and proposed parks and playgrounds.

The primary function of such an official map is to implement the community's master plan of highways by, in essence, prohibiting the construction of new buildings in the mapped beds of future streets, as well as in the mapped beds of partially or wholly developed streets that are to be widened. A secondary function of the official map is to similarly implement the community's master plan of parks and open spaces; and in this respect it can be used to protect scenic and historic sites, watercourses and drainageways, and floodplains and marshes. *An incidental, but important, benefit accruing to the community through properly executed official mapping is, of course, the stabilization of the location of real-property boundary lines both private and public.*

Insofar as the official map allows the municipality to reserve land for public purposes without commitment to actual purchase, it functions as a refinement of the community's master plan, reflecting certain aspects thereof in a precise, accurate, and legally binding manner. On completion of the topographic and property boundary-line base maps described herein, specific projects—such as new major streets and highways, proposed street widenings, relocations, vacations, proposed parks, parkways, or drainageways—may be taken from the master plan, detailed as to specific location, placed on the base maps, and the base maps adopted as portions of the community's official map. Thus, by exercise of the police power, property boundary lines can be stabilized and positive direction given to future community development.

---

*Other sections of the statutes make the official map act applicable to villages and towns as well as cities.

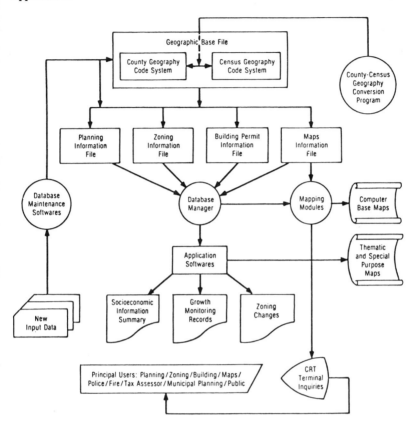

FIGURE A.9   Conceptual design of DuPage Development Department data base.

COSTS

Costs of creating the described foundation for a multipurpose cadastre are recorded in four categories by the Commission: (1) land surveys, (2) control surveys, (3) topographic map compilation, and (4) cadastral map compilation. In any consideration of such costs, the complexity of the factors influencing the unit costs must be recognized, including particularly the size, configuration, and character of the area to be mapped. Based on the experience in southeastern Wisconsin, the costs of the land surveying entailed in relocating and monumenting the U.S. Public Land Survey corners, establishing the witness marks and related ties, and preparing the necessary dossier sheets and attendant certificates throughout the areas mapped has ranged from a low of about $60 per corner to a high of about $400 per corner, typically approximating $200 per corner in 1980 dollars. The control-survey costs, including establishment of state plane coordinates and elevations for the monumented U.S. Public Land Survey corners, has ranged from a low of about $70 per corner to a high of about $660 per corner, typically approximating $400 per corner in 1980 dollars. Topographic map-compilation costs have ranged from a low of about $580 to a high of about $8200 per square mile, typically approximating about $1700 per square mile for the 1:2400-scale mapping and $4200 per square mile for the 1:1200-scale mapping, including photography. Costs of preparing the cadastral maps have ranged from $10 to $40 per parcel, typically approximating $10 per parcel.

SUMMARY AND CONCLUSION

Properly designed and executed, a good local mapping program can provide the geographic reference framework, base maps, and cadastral overlay that constitute three of the fundamental technical requirements for the creation of a sound multipurpose cadastre and for its eventual expansion into a multipurpose land-data bank.

The geometric framework for the spatial reference of data is one of the most important factors on which the ultimate success or failure of any multipurpose cadastre and land data bank will depend. The necessary geometric framework must permit identification of land areas by coordinates down to the individual parcel level while permitting the precise mathematical correlation of real-property boundary and earth-science data. This requires the relocation and monumentation of all the U.S. Public Land Survey corners within the geographic area for which the land data system is to be created and the utilization of these corners in stations of high-order traverse and level nets tied to the National Geodetic Datum. The traverse nets establish the true geographic positions of the U.S. Public Land Survey corners in the form of state plane coordinates, thereby providing a common geometric framework for the collection and coordination of both cadastral and earth-science data, as described in Section 6.3 of this report. The monumented, coordinated corners in turn provide the basis for readily maintaining the data base in current conditions since all future surveys can be readily tied to these corners.

The specific geometric framework used in southeastern Wisconsin is, of course, applicable only to those parts of the United States that have been covered by the U.S. Public Land Survey System. The fundamental concept involved—the need to place both cadastral and earth-science data on a common geometric base—is, however, applicable to any area. In those portions of the United States that have not been covered by the U.S. Public Land Survey System, the application of this concept may well be more difficult and costly, requiring the incremental placement on the State Plane Coordinate System of some special network of survey control points adapted to each locality. Nevertheless, this work is just as essential if a comprehensive land-data system is to be created over time. Once the geometric framework is in place, the preparation of the base map necessary to record the basic geography of the area and the preparation of the cadastral overlay to the base map become relatively simple operations. The preparation of such maps is, moreover, essential to sound community development and redevelopment, a fact that should dictate the preparation of such maps in any case.

The importance of the establishment of a sound geometric framework and related maps as a sound foundation for multipurpose cadastres and land-data banks is apt to be overlooked by decision makers as a technical detail in their deliberations over the other important issues involved in the creation of such systems. The establishment of a sound geometric framework, and the proper preparation of the related base maps and cadastral overlays is, however, a fundamental undertaking that clearly will require much understanding, foresight, and commitment on the part of the technicians and decision makers concerned. Failure to make the proper decisions concerning the basic technical foundation of any land-data system during its formative period will jeopardize the future utility of the system, for reform will become increasingly costly and difficult over time.

Appendix A.2
DuPage County, Illinois

INTRODUCTION

In late 1979 the County Board of DuPage County, Illinois, acted to create a county remonumentation and integrated computer-mapping program. In taking this action, the County Board recognized the high costs and inefficiencies attendant to inadequate survey control and mapping and the need to meet in a more consistent, coordinated, and cost-effective manner the control survey data and mapping requirements of the literally hundreds of government and private agencies operating within the county, utilizing the latest state-of-the-art technologies. The Board noted that cooperation on joint projects between county agencies had often been hindered by the inability of the departments to readily utilize each other's data. In many instances, various levels and agencies of government, as well as private organizations, within the county were duplicating survey data or maps that already existed in some other department or agency files.

DuPage County is located in the Chicago metropolitan area. It has an area of about 340 square miles and a population of over 650,000 persons and is one of the fastest-growing counties in the United States. It is subdivided into over 250,000 parcels of land. All property descriptions in the county are based on the U.S. Public Land Survey System. As of 1979, however, only a very few of the approximately 1400 public-land survey corners in the county were permanently monumented. In many instances, conflicting locations for these corners resulted in gaps or overlaps in land-ownership descriptions and related uncertainties of title. Lack of geodetic survey control with respect to the section and quarter-section corners and attendant inability to relate real-property boundary-line data to a map projection made it next to impossible to compile an accurate cadastral map of the county.

DESIGN OF THE PROGRAM

The responsibility for implementing the program established by the County Board was assigned to the Supervisor of the Assessment, Maps, and Plats Division of the DuPage County Development Department. Recognizing the need for an overall plan to guide the execution of the County Board-mandated program, the Department retained a consultant with expertise in the field of cadastral and control surveys to help design and implement the control-survey and computer-mapping system. During the design phase of the program, three major goals were identified:

1. All public-land survey corners in the county are to be restored and monumented. The land survey work involved is to be done by local land surveyors retained for this purpose, working under the direction of the overall consultant. The corners are to be marked by cast iron monuments, and all the monumented corners are to be tied to the National Geodetic Survey Control Network and State Plane Coordinate values established for the corners.

2. The monumented survey control network is then to be used as a geometric framework for the development of a comprehensive mapping system that will meet the needs of all potential users. The maps are to be produced in digital form for computer manipulation so that map data could be readily produced at any scale or format desired. The major hardware components of the computer-mapping system will include a county-owned large mainframe computer and mass data storage units, input stations consisting of cathode-ray-tube stations and digitizing tablets, and output stations consisting of plotters and line printers and attendant software programs.

3. The computer-mapping system is to be designed to interface with all county records also in the process of being converted to computer-readable form. Such interfaces will permit the linkage of zoning, building permit, and land-use information, as well as land-title record and tax-assessment information, to real-property parcels by machine.

INERTIAL SURVEY

To provide initial survey control both for the preparation of the orthophoto maps and for the subsequent coordination of the U.S. Public Land Survey corners, the National Geodetic Survey Control Network was densified utilizing inertial survey techniques. The inertial survey was designed to establish a horizontal control station every 2 miles throughout the county. Approximately 290 miles of inertial traverse were required to establish the horizontal and vertical positions of 53 control survey stations. The inertial traverse work was supplemented by an additional 21 miles of ground traverse to ensure that all the stations met second order, class II, standards. The total cost of this inertial control densification was approximately $40,000, or about $755 per station.

U.S. PUBLIC LAND SURVEY CORNER REMONUMENTATION

As of mid-1982, 185 public-land survey corners, including centers of sections, have been permanently monumented and tied into the State Plane Coordinate System. The remonumentation of the public-land survey corners is proceeding in parallel with the preparation of cadastral overlays for the previously prepared orthophotographs of each survey township in the county. The schedule calls for two survey townships to be completed each year, with the program being completed by the end of 1986.

The public-land survey corners are tied to the State Plane Coordinate System by conventional ground traverse. The cost of corner relocation and monumentation has approximated $825 per corner. The cost of establishing state plane coordinates and elevations related to the National Geodetic Datum has approximated $265 per corner.

ORTHOPHOTO BASE MAPS

Base maps are being prepared in a staged program in conjunction with the schedule of the remonumentation program; each map will cover one U.S. Public Land Survey section within the county. These base maps consist of 1:24,000-scale orthophotographs prepared by a photogrammetric engineer. The specifications governing the preparation of the orthophotographs required that the Illinois State Plane Coordinate System be marked on the photographs at 1000-foot intervals to National Map Accuracy Standards. To achieve the desired accuracy, control points were paneled prior to photography for later use in analytical aerotriangulation procedures. The preparation of orthophoto maps has cost $400 per square mile, while the cost of preparing the cadastral overlay to the orthophoto base maps has cost $2.00 per parcel, or approximately $1500 per square mile.

DIGITAL MAP SYSTEM

The computer-mapping system of DuPage County has been designed to meet the most demanding accuracy requirements, that is, for engineering plans. The remapping of each section will involve correlating all the survey data developed through the remonumentation program with existing plats of record.

In the spring of 1981 the county purchased the basic hardware for the system: a Tektronix 4054 graphics workstation (19-inch CRT, dual-disk drives, and 36-inch × 48-inch tablet), and a Calcomp 970 plotter. The graphics workstation can operate either as a stand-alone computer, programmable in BASIC, or as a terminal of the county's mainframe computer. The mapping data compiled through December 1982 will reside on the mainframe computer of the software consultant, to be installed on the county computer early in 1983.

DuPage County has completed about half of its plan to automate the land-data functions of the county government. Within a few years, daily operations such as building permits, land-use studies, and zoning enforcement all will be processed on the county's computer.

TOTAL SYSTEM DESIGN

The DuPage County program is unusual, if not unique, within the United States in that it is proposing to create a multipurpose cadastre in a single operation spanning

TABLE A.1 Estimated Costs and Cost Savings of the DuPage County Computer-Mapping and Remonumentation Program (All Figures Computed as 1982 Dollars.)[a]

Costs of the Remonumentation/Computer-Mapping System, 1980-1985	
Remonumentation, 1983-1985	$ 475,000
Computer hardware, 1981-1982	170,000
Computer software, 1982	160,000
Computer software maintenance, 1983-1985	150,000
Contractors and other, 1980-1982	334,058
TOTAL	$ 1,289,058
Annual Cost Savings, Starting in 1985	
Estimated at one half of the following present mapping costs:	
DUPAGE COUNTY GOVERNMENT	
County Clerk	$ 320,000
Maps and Plats Division	100,000
Public Works and Highway	80,000
Planning, Building, and Zoning Divisions	75,000
Forest Preserve District	40,000
Election Commission	35,000
Supervisor of Assessment	30,000
Health Department	20,000
Superintendent of Schools	5,000
MUNICIPALITIES	
34 cities and villages	
@ $20,000 each	680,000
LOCAL TAX BODIES	
Sanitary Districts and Fire Protection Districts	100,000
TOTAL ESTIMATED ANNUAL MAPPING COSTS	$ 1,485,000
ESTIMATED ANNUAL COST SAVINGS (ONE HALF)	$ 742,500

[a]Source: Donahue (1982).

a period of about 7 years, including not only the provision of survey control, base maps, and cadastral overlays but also the digitization of the maps and the creation of a file permitting the correlation of all land-related county records by machine processing. The conceptual design of the system is indicated in Figure A.9. The project design consultant has provided estimates of annual cost savings in county and municipal mapping costs that amount to 58 percent of the expenditures required to complete the system, during the period 1980-1985, as listed in Table A.1. Overall, the DuPage County system is being implemented in substantial accord with the principles set forth in this report, with the promise of significant economies in the operation of key county departments over the years ahead.

Appendix A.3
Jefferson County, Colorado

INTRODUCTION

In 1977 the Commissioners of Jefferson County, Colorado, acted to create a Mapping Department and to charge that department with the responsibility of producing and maintaining accurate, up-to-date, large-scale maps of the county. In this action, the Commissioners recognized the importance of good maps to the planning for, and management of, a rapidly growing county that is an integral part of the five-county Denver metropolitan area. In the relatively short time since its creation, the Mapping Department has made significant progress toward meeting its charge of accurately mapping the county. The work of the department has consisted of a logical progression of steps leading in 1982 to the threshold of a computer-aided mapping capability and the creation of a digital geographic data base and has provided a foundation for the eventual creation of a multipurpose cadastre.

The major components of the geographic data base are an ongoing mapping program, a computer-aided geographic locator system, and a county data-processing system used by the County Assessor, the County Clerk and Recorder, and the County Planning and Engineering Departments. This case study focuses on the mapping program, which provides an important technical basis for the data base, describing the important components of that program.

SURVEY CONTROL

Even prior to the creation of the County Mapping Department, responsible county officials recognized the importance of good survey control to any mapping program.

155

Accordingly, the county and the National Geodetic Survey entered into an agreement to undertake a joint project to densify the basic first-order horizontal control network in the county. Prior to 1976 only three first-order horizontal control stations existed in the 780-square-mile county, all located on relatively inaccessible mountain peaks. Under the joint program, National Geodetic Survey crews established 51 first-order horizontal survey control stations within the county. The stations were marked by standard National Geodetic Survey concrete monuments with brass caps, and each monument was provided with reference and azimuth marks, as required by the National Geodetic Survey practice. The horizontal position of the stations, in terms of their latitude and longitude and corresponding state plane coordinates, were established by first-order triangulation surveys conducted to National Geodetic Survey Standards. The completed first-order triangulation network provides monumented stations at about a 5-mile spacing thoughout the county. The cost of the completed first-order triangulation network was $1350 per station.

County survey crews trained by the National Geodetic Survey then established an additional 100 secondary horizontal survey control stations within the county. These stations were marked by concrete monuments with brass caps; and the positions of these stations in terms of state plane coordinates were established by traverse surveys conducted to National Geodetic Survey Standards for second-order, class I, traverse surveys. The cost of this work was $500 per station. The second-order stations, together with the first-order stations, thus provided a control survey station spacing of about 2 miles throughout the county.

Subsequent to the establishment of these 151 first- and second-order horizontal control survey stations within the county, the county land development regulations were amended to require that all new land subdivisions be tied to the geodetic control survey network. The new regulations required that the ties be made by third-order, class II, traverse surveys. These regulations thus provide for the establishment of state plane coordinates on property boundary lines down to the parcel level. Since Colorado law requires that all subdivision plats also be tied to the U.S. Public Land Survey System, this provision in the county land development regulations also results in the establishment of state plane coordinates for section and quarter-section corners as a part of the land subdivision process. In the 3^1/$_2$ years in which the survey tie requirement has been in effect, 295 U.S. Public Land Survey section and quarter-section corners have been recovered and placed on the State Plane Coordinate System. The cost of this work has been borne by the private developers involved in the land subdivision process.

Additional local cadastral survey control points are established on the boundaries of all subdivisions approved by the county, which are required to be monumented. The private surveyors who locate these monuments have been very supportive of the objectives of the county cadastral survey program, according to the staff of the Mapping Department, and have voluntarily submitted photocopies of their field notes. The modest additional costs of maintaining high standards of survey control, monumentation, and documentation for the new subdivisions are passed on to the land developer and, eventually, to the buyers of the individual parcels.

ORTHOPHOTOGRAPHY

Following the completion of the first- and second-order survey control densification, a contract was let with a private photogrammetric engineering firm to produce orthophotography of the county. In the densification program, 110 of the first- and second-order horizontal control stations established were paneled; and in October 1976 aerial photography was obtained at a negative scale of 1 inch = 1000 ft. This photography was then processed into orthophotography on dimensionally stable base material at a scale of 1:4800, the orthophotography being referenced to the Colorado State Plane Coordinate grid system. Each orthophotograph covers an 8000 ft × 11,000 ft area so that 271 orthophotographs would cover the entire county. The county maintains this 1:4800 base-map system for 214 of these areas, excluding the southern one fifth of the county, which falls within the Pike National Forest.

The orthophotographs are updated using new aerial photography flown every 4 years. As section and quarter-section corners are recovered and tied into the State Plane Coordinate System, their locations are plotted on the orthophotographs. These orthophotographs form the basis for the preparation of all county maps. The preparation of initial orthophotographs cost $60 per square mile; subsequent updating of the orthophotographs cost $12 per square mile.

PLANIMETRIC MAP SERIES

Subsequent to the completion of the orthophotographs, the Mapping Department undertook the preparation of a planimetric base-map series for the county. The maps are produced by drafting in ink on pin-registered dimensionally stable base material overlays to the 1:4800-scale orthophotographs. Overlays to these planimetric maps show the locations of all the primary and secondary horizontal survey control stations, the locations of such section and quarter-section corners as have been recovered and placed on the State Plane Coordinate System; all platted subdivisions; all streets, highways, and railroads; and major topographic and hydrographic features. Certain source documents, including subdivision plats, street and highway right-of-way maps, and survey records are used in the compilation of the planimetric maps.

Each map sheet covers the same area as the corresponding orthophotography. The cost of the finished planimetric maps is $375 per square mile, exclusive of control and orthophoto costs.

However, the overlay of subdivisions is not yet a complete cadastral overlay because:

1. Standard parcel identifiers are not indicated on the subdivided parcels. Most county records are indexed according to the assessor's block and lot numbers, but other maps must be consulted to determine the number of a given parcel.

2. The county has not had the resources to map property boundaries in the "aliquot lands" comprising most of the land area of the county, which to date have been subdivided with reference to the lines of Public Land Survey System (PLSS) sections and fractions of sections and not under subdivision plans approved by the county.

The latter situation underlines the need for locating all the corners and quarter-corners of the PLSS sections by state plane coordinates, so that this predominating system of rural property boundary markers may be spatially related to the county property map system.

CURRENTLY PLANNED SYSTEM EXTENSION

As of February 1982, Jefferson County has retained a licensed land surveyor to recover U.S. Public Land Survey section and quarter-section corners and to tie those corners to the geodetic control network in a systematic manner in order to provide a tertiary control network throughout the county. The Public Land Survey corners are to be tied to the National Geodetic Survey Network by traverse surveys conducted to third-order, class I, standards. The resulting data are to be stored in the form of state plane grid coordinates in the county computer data bank and ultimately provide the framework for a digital mapping system.

A proposal has been submitted to the Jefferson County Commissioners for an entry-level computer-mapping system consisting of a desk-top graphics computer (with 64 kbyte of memory, expandable to 4 Mbyte), 20 Mbyte hard-disk systems, line printer, digitizer, and plotter. This system will expand the countywide Geographic-Information System, utilizing both digitized and engineering data, as an integral part of an overall Management-Information System.

Appendix A.4
The Philadelphia Area

INTRODUCTION

The Regional Mapping and Land Records (RMLR) program has been developing and testing a computerized system of large-scale digital mapping, joined to a parcel-based land-records data base, to serve the Delaware Valley region of Pennsylvania and New Jersey, which is centered in the city of Philadelphia. The program was organized in 1976 through the efforts of key individuals in local and regional government, utilities, and other private-sector interests, who had long worked from various perspectives with the inadequacies of existing land-records and mapping systems. In initiating the RMLR program the participating agencies recognized a number of common issues: the present mapping and data systems being separately maintained were obsolete and costly; there was significant duplication of effort in producing parallel sets of data and maps; new technology was available to produce more efficient systems; and the start-up cost of a new system made it prohibitively expensive if borne solely by a single user.

The principal elements of the RMLR system are (1) computerized digital maps produced at a variety of scales, jointly produced and maintained by a consortium of public and private agencies for the common use of all participants; (2) a federated data base consisting of a wide range of information files, produced and maintained individually by the participants and shared at the discretion of the owners; (3) co-ordination of the local map and data base with state and national programs to increase mutual support and eliminate duplication (Delaware Valley Regional Planning Commission, 1980).

With the present land-records and mapping system, each change in the Philadelphia street network must be updated 28 times in city agencies alone. A normal 50 changes each year results in 1400 changes in the street network maps. In addition, the data associated with these maps must also be updated. Similar redundancy has been experienced by each of the utilities in the region, as well as the title insurance industry.

There are over 650,000 properties in the city of Philadelphia. Another inefficiency occurs because of the lack of an uniform property identifier in the city and the physical separation of related land records. Sample information searches have revealed that it takes over 3 hours, excluding travel time, to find ownership, mortgage, lien, violation, zoning permit, land-use, assessment, and tax-revenue information on a specific property (Hadalski, 1982).

RMLR I PILOT PROJECT

In order to test the concepts of the proposed map system, RMLR conducted a pilot project within the Philadelphia metropolitan region. The specific objectives of the project were to test the technical methods and operating procedures proposed and to determine the costs and benefits of the proposed system. A 50-square-mile region in Montgomery County, in and around Norristown, Pennsylvania, was selected for the pilot project site. This area provided an excellent mixture of rural, suburban, and high-density urban land use, typical of most urbanized areas of the region. In addition, existing horizontal and vertical ground control was distributed in such a fashion as to be almost ideal for the project.

The pilot project design was completed in December 1976. A contract was negotiated in October 1977, and, after some last-minute fund raising, work began in February 1978. The direct RMLR budget was $100,000, with $75,000 being used for the pilot project contractor consultant, Vernon Graphics. The project steering committee was formed consisting of those utilities and governments making financial contributions. They included Philadelphia Electric Company; Bell of Pennsylvania; Philadelphia Gas Works; Pennsylvania Power and Light; Philadelphia Suburban Water Company; Montgomery County, Pennsylvania; City of Philadelphia; Bucks County, Pennsylvania; Chester County, Pennsylvania; Delaware County, Pennsylvania; the U.S. Army Corps of Engineers; and the Delaware Valley Regional Planning Commission. Advisory members to the steering committee included the Pennsylvania Land Title Association, the National Geodetic Survey, the U.S. Geological Survey, and other state and federal government agencies.

In planning for a regional mapping and data system, which would provide the base on which to build a comprehensive land-record system, the steering committee decided that the basic geography, which would include road cartways and center lines, drainage, bridges, and railroads, should be of the highest degree of accuracy attainable under state-of-the-art technology. The system would be designed in such a fashion as to provide for all foreseeable future accuracy requirements. This resulted in the design of testing in the project to determine the scale of aerial photography needed to produce required degrees of accuracy, to evaluate the relative merits of various methods of photoanalytical aerotriangulation, and to determine the feasibility of direct stereo digitization (Delaware Valley Regional Planning Commission, 1980).

BASE MAPPING IN THE RMLR I PILOT PROJECT

Aerial photography was obtained at three different scales: 1:16,000, 1:8000, and 1:4000. The 1:16,000-scale photography was used for rural and suburban orthophotography and digital mapping at scales of 1:4000 and 1:2000. The 1:8000-scale photography was used for suburban and urban orthophotography, and digital mapping at a scale of 1:1000, and the largest-scale 1:4000 photography was used for urban digital mapping at a scale of 1:500.

All horizontal and vertical ground-control points were established and paneled prior to the taking of the aerial photography. In establishing supplemental control using analytical aerotriangulation for the purpose of leveling and scaling each photogrammetric stereomodel, two methods of adjustment were used: polynomial adjustment and bundle-block adjustment. Field surveys were conducted to check the results of each aerotriangulation adjustment method. Overall, the results of the bundle adjustment proved to be 20 percent more accurate than those developed by the polynomial method.

The orthophotographic negatives were compiled at four times the original photographic negative scale and, if required, subsequently enlarged to eight times the original scale with satisfactory quality. The digital-mapping data base was produced using direct stereo digitization, digitizing directly from the aerial photographic stereo model using stereo compilation equipment, with a direct link to the computer system. The project was designed to meet National Map Accuracy Standards. Digitized features included base-map features (roads, drainage, and railroads), contours, utility poles, utility transmission towers, manholes, fire hydrants, culverts, utility buildings, transformer pads, utility fences, gas-regulator boxes, generating-station facilities, pipeline traces, sewage disposal and waste-treatment plants, pumping stations, radio and microwave towers, and storage tanks. All digitized detail was reviewed on cathode-ray-tube screens as digitized and later at a second station for cleanup and enhancement of detail. Names, text, and proper identities were added during this phase for all features. Edit plotbacks of each map were made on a flatbed plotter in ballpoint on Mylar, with color separation to enhance readability. These plotbacks were reviewed for changes, additions, revisions, and corrections as required (Delaware Valley Regional Planning Commission, 1980.)

COSTS OF MAPPING IN RMLR I

The RMLR Steering Committee advises that any attempt to generalize from the costs experience described in this section should recognize the following: (1) the value of a multipurpose, multijurisdictional land-data system will be unique to the locality for which it was designed; (2) local administrative, managerial, and political conditions will present obstacles that are at least as difficult to overcome as the cost of the technology; (3) RMLR I was a demonstration of a variety of alternative procedures, not all of which will be implemented in subsequent operational phases of the project; and (4) the range of local conditions covered in RMLR I may have been too narrow a sample to be representative of the state of the art, especially since the "learning curves" affected the productivity of both the vendor and the users.

Table A.2 summarizes the costs of preparation of base maps in the RMLR I pilot project in 1978-1979 dollars, per square mile. The digital base maps include streets and alleys, drainage, railroads, buildings, and bridges (Hadalski, 1982).

Maps of cadastral parcels were digitized in the RMLRI Pilot Project in 1978 and 1979, testing two alternative procedures that varied in cost and capabilities of the digitizing systems and in the scope of the map files created. The more elegant procedure involved intelligent graphics terminals with software that computed selected dimensions of each parcel (such as frontage) and related them to the corresponding entries in the assessor's file. The less-expensive procedure provided only limited interactivity, equivalent to automated drafting, but used the same data-base man-

TABLE A.2 Costs of Production of Base Maps in the RMLR I Pilot Project (1978-1979 Dollars per Square Mile)

Type of Area	Rural	Suburban			Urban
Map scale	1:4,000	1:2,000	1:1,000	1:1,000	1:500
Aerial photography scale	1:16,000	1:16,000	1:8,000	1:8,000	1:4,000
ORTHOPHOTO BASE-MAP COSTS[a]	$255	$385	$1394	$1394	None at this rate
Direct stereo digitization (excluding editing and audit corrections)	$425	$575	$1144	$1950	$3015
TOTAL COSTS FOR DIGITIZED BASE MAPS AND ORTHOPHOTOS	$680	$933	$2538	$3344	$4314 without orthophotos

[a]Includes aerial photography, photoanalytics, and aerotriangulation according to the fully calibrated bundle-adjustment method and orthophotography producing screened positives.

TABLE A.3   Costs of Alternative Cadastral Mapping Procedures in the
RMLR I Pilot Project (1978-1979 Dollars per Parcel)

Type of Area:	Rural	Suburban	Urban
Map Scale:	1:4000	1:1000	1:500
A. With intelligent terminals	$15.50	$6.00	$8.30
B. With entry of graphics only	11.10	3.00	3.90

agement system. The costs of producing the digital cadastral overlays, per parcel, are summarized in Table A.3. These costs include parcel layout, parcel data entry, digitizing, ballpoint plotbacks, edit corrections, final ink plots, and supervision (Hadalski, 1982).

These cost figures clearly indicate that the highly automated RMLR system is cost-effective when compared with less-sophisticated conventional mapping. This will be increasingly so through the 1980's as computer costs decline and the need for automation in data handling increases (Delaware Valley Regional Planning Commission, 1980).

PARCEL MAPPING IN RMLR I

To devise a Parcel-Information System that would better serve the needs of the assessor than existing property maps, digital property maps were prepared as a demonstration for 10 square miles of the RMLR I test area. Existing tax maps were used as much as possible as the source material for the digitization of parcels and their associated data. Digitized base-mapping elements were utilized in conjunction with orthophotography in the tax-parcel layout, ensuring complete compatibility of all features in the geographic and property map bases (Delaware Valley Regional Planning Commission, 1980).

The Recorder of Deeds and the Assessor of Montgomery County, Pennsylvania, who have been hosts to the RMLR I pilot project, have been stimulated by the RMLR program to organize an integrated land-parcel data base that now serves both offices and is being adopted by private title insurance companies as the definitive index of land parcels. The offices that maintain the cadastral parcel indexes and maps were brought together on the same floor with the deed recorder in the county building in Norristown, so that new parcel identifiers can be activated almost immediately when the documents creating the parcels are filed. The establishment of a parcel index for all land-title documents being filed at the Recorder's Office was found to be within his existing statutory authority.

DATA PROCESSING IN RMLR I

The systems software and applications software developed in the RMLR I pilot project provide the control for the entry management, analysis, and output of both graphical and alphanumeric data. Multilayering of data types provides for access and retrieval of data by defined output map limits or name, categories of elements, x-y window limits, and coordinate search within a defined category. It is thus possible to choose an area of any size and shape and to request a variety of graphics and associated data. For example, one can request all the roads within a township in association with schools, firehouses, and police stations. Specific types of data, such as all parcels within a given area that fall within 100-year floodplains or that are assessed at between $25,000 and $50,000, or any types of data that can be oriented geographically, can be retrieved similarly (Delaware Valley Regional Planning Commission, 1980).

SUMMARY OF RMLR I OBJECTIVES AND RESULTS

1. *Objective*:   Test the technology of large-scale digital mapping.
   Results:
   - Large-scale direct stereo digitization does work well and can meet high accuracy standards.
   - The capacity of interactive graphics to manipulate and manage public files of geographic data is substantial.
   - At appropriate scales, a high rate of interpretability of features from aerial photography, including utilities, is very evident.

2. *Objective*:   Develop detailed estimates of the costs of implementing a computer graphics system.
   Results:
   - The RMLR I pilot project developed and analyzed costs for common elements and specific elements that provide a basis for planning a system.

3. *Objective*:   Develop measures of the uses and benefits of a RMLR system.
   Results:
   - RMLR I was able to identify many uses and applications of a more precise mapping system that covers the utilities, county and city functions, management, and graphics and data retrieval of a variety of information.

4. *Objective*:   Develop a strategy for implementing a modern mapping system.
   Results:
   - The RMLR I experience confirms that no major problems exist of a technical

nature. Rather, the problems may be institutional and therefore require involvement of technical, administrative, and financial management of a potential user organization or group (Delaware Valley Regional Planning Commission, 1980).

THE RMLR II PILOT PROJECT

To further demonstrate and test the technology developed for the RMLR I pilot project, especially its use in a dense central-city environment, a second pilot project has been initiated. RMLR II will cover an area 5 blocks square (0.27 square mile) within center-city Philadelphia that contains the tallest buildings in the region along with the most concentrated complexes of land use and subsurface infrastructure. It is part of the oldest survey district in the city, where property records are based on very old plans, some of which date from the late 1600's. RMLR II is planned to develop updating and maintenance procedures, transmission procedures, and workstation support for further applications of the RMLR system. It will be handling special problems of ground control, paneling, aerial photo annotation, and stereo digitizing that were not encountered in the RMLR I project area. The project is being augmented by special contracts for individual agencies within RMLR to develop (1) special software routines, including a DIME to RMLR interface; (2) specialized workstation training; and (3) "scanning" as a separate procedure to enter property boundary data into the digital data base.

Aerial photographic coverage of a larger area of 1.72 square miles, including center-city Philadelphia, was flown in mid-July 1982 at scales of 1:8000 and 1:4000. Fifteen existing ground-control points were available to control analytical aerotriangulation using the photographic coverage. The planimetric base maps are to be stereo digitized at a scale of 1:500 according to National Map Accuracy Standards. The planimetric features are being digitized in separate data layers to permit computer plotting of any individual data layer or any combination of data layers. The following data categories are being used:

Basic Map Features (cartways, traffic islands)	Telephone Booths
Street Light Poles	Radio, TV, Microwave Towers
Utility Poles	Bus Shelters
Traffic Light Standards	Trees
Manholes	Subway Entrances
Fire Hydrants	Subway Vents
Culverts	Subway Lights
Catch Basins	Bollards
Steam and Gas Vents	Serving Area Connectors (Telephones)

Buildings	Transportation Facilities
Fire Stations	Other
Police Stations	Parks
Hospitals and Health Centers	Signposts
Educational Centers	Street Centerlines and Inter-
Historical Buildings	sections
Municipal Buildings	Service Vents
Utility Buildings	

For each data category in which names are associated with a data item, such as streets or buildings, a data base of names will be kept for retrieval and plotting. The planimetric digital data base will allow the flexibility of plotting with reference to any of the following three horizontal reference datums: South Zone of the Pennsylvania State Plane Coordinate System, latitude and longitude, or Zone 18 of the Universal Transverse Mercator (UTM) 6° Grid. The system will provide for computer plotting of any specific portion of the data base at any specified scale in any combination of data layers (Smith, 1982).

Within the digital planimetric base map area of the RMLR II pilot project there exist approximately 4300 deeded properties, including over 2000 parcels, 2000 condominium units, and 13 airspace rights. Twenty-two tax plats are associated with the area. Using the computer-plotted basic map features and buildings as a base map for new tax plats, the city of Philadelphia Records Department will be adding parcel boundary lines, street right-of-way lines, and associated text. The parcel boundary lines will then be digitized into a digital data layer, parcel centroids determined, and parcel-specific information added to the data base from the files of the city of Philadelphia Board of Revision of Taxes, to be merged with the information from the Records Department.

In order to ensure that the digital data tapes will be compatible with the operating systems of the major participants in the project, IBM IGGS Interface Format has been specified. This will permit greater in-house use of the generated digital data in an operational environment, for additional testing and procedure validation.

By maintaining detailed cost records of each phase of the RMLR II pilot project, the RMLR steering committee will have a firm base for the formulating of future project decisions.

With the awarding of the RMLR II contract, the city of Philadelphia, Philadelphia Electric Company, Philadelphia Gas Works, and Southeastern Pennsylvania Transportation Authority (SEPTA), along with the Delaware Valley Regional Planning Commission are continuing their commitment to go forward together, at a cautious pace, in building a modern, large-scale digital-mapping system. These leaders of the RMLR program appear to have developed a successful formula for joint sponsorship and use of the mapping system by the local government and the major utilities, which has eluded so many others.

Appendix B

United States National Map Accuracy Standards

With a view to the utmost economy and expedition in producing maps which fulfill not only the broad needs for standard or principal maps, but also the reasonable particular needs of individual agencies, standards of accuracy for published maps are defined as follows:

1. **Horizontal accuracy.** For maps on publication scales larger than 1:20,000, not more than 10 percent of the points tested shall be in error by more than 1/30 inch, measured on the publication scale; for maps on publication scales of 1:20,000 or smaller, 1/50 inch. These limits of accuracy shall apply in all cases to positions of well-defined points only. Well-defined points are those that are easily visible or recoverable on the ground, such as the following: monuments or markers, such as bench marks, property boundary monuments; intersections of roads, railroads, etc.; corners of large buildings or structures (or center points of small buildings); etc. In general what is well defined will also be determined by what is plottable on the scale of the map within 1/100 inch. Thus while the intersection of two road or property lines meeting at right angles would come within a sensible interpretation, identification of the intersection of such lines meeting at an acute angle would obviously not be practicable within 1/100 inch. Similarly, features not identifiable upon the ground within close limits are not to be considered as test points within the limits quoted, even though their positions may be scaled closely upon the map. In this class would come timber lines, soil boundaries, etc.
2. **Vertical accuracy,** as applied to contour maps on all publication scales, shall be such that not more than 10 percent of the elevations tested shall be in error more than one-half the contour interval. In checking elevations taken from the map, the apparent vertical error may be decreased by assuming a horizontal displacement within the permissible horizontal error for a map of that scale.
3. **The accuracy of any map may be tested** by comparing the positions of points whose locations or elevations are shown upon it with corresponding positions as determined by surveys of a higher accuracy. Tests shall be made by the producing agency, which shall also determine which of its maps are to be tested, and the extent of such testing.
4. **Published maps meeting these accuracy requirements** shall note this fact on their legends, as follows: "This map complies with National Map Accuracy Standards."
5. **Published maps whose errors exceed those aforestated** shall omit from their legends all mention of standard accuracy.
6. **When a published map is a considerable enlargement** of a map drawing (manuscript) or of a published map, that fact shall be stated in the legend. For example, "This map is an enlargement of a 1:20,000-scale map drawing," or "This map is an enlargement of a 1:24,000-scale published map."
7. **To facilitate ready interchange and use of basic information for map construction** among all Federal mapmaking agencies, manuscript maps and published maps, wherever economically feasible and consistent with the uses to which the map is to be put, shall conform to latitude and longitude boundaries, being 15 minutes of latitude and longitude, or 7.5 minutes, or 3–3/4 minutes in size.

Issued June 10, 1941   **U.S. BUREAU OF THE BUDGET**
Revised April 26, 1943
Revised June 17, 1947

Morris M. Thompson, Maps for America: Cartographic Products of the U.S. Geological Survey and others, 1979, p. 104.

References

Almy, R. R., Current land record systems in the U.S., in *Monitoring Foreign Ownership of U.S. Real Estate, Vol. 2*, U.S. Dept. of Agriculture, Washington, D.C. (1979a).

Almy, R. R., The joint development and use of property information, *Assessors J. 14*, 73-92 (June 1979b).

American Planning Association, *Planning*, American Planning Association, Chicago, Ill. (October 1981).

American Public Works Association Research Foundation, *Procurement Specifications for an Interactive Graphics System, CAMRAS Manual Part 2*, Chicago, Ill. (1979a).

American Public Works Association Research Foundation, *File Format for Data Exchange between Graphic Data Base, CAMRAS Manual Part 3*, Chicago, Ill. (1979b).

American Public Works Association Research Foundation, *Guidelines for Systems Analysis of User Requirements, CAMRAS Manual, Part 4*, Chicago, Ill. (1981a).

American Public Works Association Research Foundation, *Guide to Procurement of CAMRAS-Type Systems, CAMRAS Manual, Part 5*, Chicago, Ill. (1981b).

American Society of Civil Engineers, Manual on Map Uses, Scales and Accuracies for Engineering and Associated Purposes, Committee on Cartographic Surveying, Surveying and Mapping Division (in preparation).

American Society of Photogrammetry, *Manual of Photogrammetry*, 4th ed., Falls Church, Va. (1980).

Archer, A. J., A Unified Approach for Mapping in Prince William County, Virginia, American Congress on Surveying and Mapping *Bulletin No. 71*, pp. 17-19 (November 1980).

Association for Computing Machinery, *ACM Guide to Computing Literature*, ACM, New York (Annual).

Auerbach Publishers, Inc., *Computers in Local Government: Urban and Regional Planning*, Pennsauken, N.J. (1980a).

Auerbach Publishers, Inc., *Computers in Local Government: Finance and Administration*, Pennsauken, N.J. (1980b).

Bernard, D., Management issues in cooperative computing, *Comput. Surv. 11*, 3-17 (March 1979).

Beuscher, J. H., and R. R. Wright, *Land Use*, West Publishing Co., St. Paul, Minn. (1969).

Brown, D. C., Analytical aerotriangulation vs. ground surveying, presented to the 1971 Semi-Annual Meeting of the American Society of Photogrammetry, San Francisco, Calif. (September 1971).

Brown, D. C., Accuracies of analytical triangulation in applications to cadastral surveying, *Surveying and Mapping 33*, No. 3 (1973).

Brown, D. C., Densification of urban geodetic data, *Photogramm. Eng. Remote Sensing 43*, No. 4 (1977).

Brown, D. C., Positioning by satellites, *Rev. Geophys. Space Phys. 17*, 199-204 (1979).

Brown, R. M., and K. Stephenson, The evaluation of purchased computer software, *Mid-South Business J. 1*, 8-11 (July 1981).

Burchell, R. W., and D. Listokin, *The Fiscal Impact Handbook: Estimating Local Costs and Revenues of Land Development*, The Center for Urban Policy Research, New Brunswick, N.J. (1978).

Bureau of Public Roads and U.S. Urban Renewal Administration, *Standard Land Use Coding Manual: A Standard System for Identifying and Coding Land Use Activities*, U.S. Govt. Printing Office, Washington, D.C. (1965), reprinted by the Federal Highway Admin., U.S. Dept. of Transportation (1977).

Chatterton, W., and J. D. McLaughlin, *Towards the Development of Modern Cadastral Standards, Proceedings of the North American Conference on Modernization of Land Data Systems (a Multi-Purpose Approach)*, pp. 69-94, N. Am. Inst. for Modernization of Land Data Systems, Washington, D.C. (1975).

Colvocoresses, A. P., Evaluation of the cartographic applications of ERTS-1 imagery, *The American Cartographer 2*(1), 5-18 (1975).

Committee on Geodesy, National Research Council, *Need for a Multipurpose Cadastre*, National Academy Press, Washington, D.C. (1980).

Committee on Geodesy, National Research Council, *Federal Surveying and Mapping: An Organizational Review*, National Academy Press, Washington, D.C. (1981).

Committee on Integrated Land Data Mapping, National Research Council, *Modernization of the Public Land Survey System*, National Academy Press, Washington, D.C. (1982).

Cook, R. N., College of Law, University of Cincinnati, Cincinnati, Ohio 45221, private communication (September 1982).

Counselman, C. C., III, The Macrometer interferometer surveyor, in *Proceedings of the Symposium on Land Information at the Local Level*, Surveying Engineering Program, University of Maine at Orono (1982).

Counselman, C. C., III, and D. H. Steinbrecher, The Macrometer: a compact radio interferometry terminal for geodesy with GPS, presented at Third International Geodetic Symposium on Satellite Doppler Positioning, Las Cruces, New Mexico (February 1982).

Dangermond, J. P., and L. K. Smith, Alternative approaches for applying GIS technology, in *Proceedings of the ASCE Specialty Conference on the Planning and Engineering Interface with a Modernized Land Data System*, Denver, Colo. (June 1980).

Degnan, J. J., Goddard Space Flight Center, Greenbelt, Md. 20771, private communication (1982).

Dekle, J. C., Selecting a computerized appraisal system, EDP 1, pp. 1-4, International Association of Assessing Officers, Chicago (November/December 1981).

Delaware Valley Regional Planning Commission, The Regional Mapping and Land Records Program—A Summary Report, Philadelphia, Pa. (July 1980).

Domenici, P. V., Legislation Relating to Land Survey Issues in the West, Congressional Record—Senate 705, pp. 52143-52146 (March 12, 1981).

Donahue, J. G., 502 Union Street, Geneva, Illinois 60134, private communication (August 1982).

Donaldson, H., *A Guide to the Successful Management of Computer Projects*, Wiley, New York (1978).

Dueker, K. J., Land resource information systems, a review of fifteen years' experience, *Geoprocessing* (1979).

Federal Geodetic Control Committee, Classification Standards of Accuracy and General Specifications of Geodetic Control Surveys, National Ocean Survey, U.S. Dept. of Commerce, Rockville, Md. (February 1974, reprinted May 1978).

Federal Geodetic Control Committee, Specifications to support classification, standards of accuracy, and general specifications of geodetic control surveys, U.S. Dept. of Commerce, Rockville, Md. (1980).

Federal Mapping Task Force, *Report on Mapping, Charting, Geodesy and Surveying*, Office of Management and Budget, U.S. Govt. Printing Office, Washington, D.C. (July 1973).

Fife, D. W., *Computer Software Management: A Primer for Project Management and Quality Control*, National Bureau of Standards, U.S. Dept. of Commerce, U.S. Govt. Printing Office, Washington, D.C. (1977).

Giles, P. B., Systems analysis and urban information, in *Developing the Municipal Organization*, S. P. Powers, F. G. Brown, and D. S. Arnold, eds., International City Management Assoc., Washington, D.C. (1974).

Hadalski, J. M., Reflections on the Regional Mapping and Land Records Project, presented at the Harvard Computer Graphics Week 1982, Cambridge, Mass. (July 1982).

Hendrix, K., Geographic positioning using various instruments and methods, in *Technical Papers of the American Congress on Surveying and Mapping*, ASP-ACSM Convention, Washington, D.C., pp. 73-82 (1981).

Henssen, J. L. G., Cadastres and land registration on the European continent, ORICRF Speech, The Hague, Netherlands (1973).

International Association of Assessing Officers, *Improving Real Property Assessment: A Reference Manual*, IAAO, Chicago, Ill. (1978).

International Association of Assessing Officers, *Standard on Property Use Codes*, IAAO, Chicago, Ill. (1981).

King, J. L., Sources of computing capability for local government: an overview, in *Computers in Local Government: Finance and Administration*, Auerbach Publishers, Inc., Pennsauken, N.J. (1980).

King, J. L., and K. L. Kraemer, Cost-benefit analysis in local government computing operations, in *Computers in Local Government: Finance and Administration*, Auerbach Publishers, Inc., Pennsauken, N.J. (1980).

Larsen, B., J. L. Clapp, A. H. Miller, B. J. Niemann, and A. L. Ziegler, Land Records: The Cost to the Citizen to Maintain the Present Land Information Base, A Case Study of Wisconsin, Department of Administration, Office of Program and Management Analysis, 64 pp., Madison, Wisconsin (1978).

Lincoln Institute of Land Policy, *National Survey of Opinion in the Attributes of a Successful Land Data System*, Lincoln Institute Monograph 82-4, Cambridge, Mass. (1982).

Lucas, J. R., Photogrammetric Control Densification Project, in *Proceedings of the Second International Symposium on Problems Related to the Redefinition of North American Geodetic Networks*, hosted by National Geodetic Survey, Rockville, Md. (April 1978).

MacDoran, P. F., D. F. Spitzmesser, and L. A. Buennagel, SERIES: Satellite Emission Range Infrared Earth Surveying, *Proceedings of the Third International Geodetic Symposium on Satellite Doppler Positioning*, Las Cruces, N.M. (February 1982).

Mancini, A., Inertial Geodesy, A Total Solution to the Geodetic Problem of Positioning, Gravity, and Deflections, Proceedings, Vol. 5, XV International Congress of Surveyors, Federation Internationale des Geometres (FIG), Stockholm, Sweden (June 1977).

Matthews, J. R., Negotiating and writing a hardware contract, Computers in Local Government: Finance and Administration, Auerbach Publishers, Inc., Pennsauken, N.J. (1980).

McLaughlin, J. D., The Nature, Design and Development of Multipurpose Cadastres, Ph.D. Thesis, U. of Wisconsin, Madison (1975).

McLaughlin, J. D., Maritime Cadastral Accuracy Study, Land Registration and Information Service Technical Report, U. of New Brunswick, Fredericton, New Brunswick, Canada (1977).

Metzger, P. W., *Managing a Programming Project*, Prentice-Hall, Englewood Cliffs, N.J. (1973).

Moellering, H., The challenge of developing a set of national digital cartographic data standards for the U.S., in Technical Papers of the American Congress on Surveying and Mapping, Washington, D.C. (1982).

Motto, J. R., Writing a request for proposal (RFP) and evaluating proposals, *Computers in Local Government: Finance and Administration*, Auerbach Publishers, Inc., Pennsauken, N.J. (1980).

Moyer, D. D., and K. P. Fisher, *Land Parcel Identifiers for Information Systems*, American Bar Foundation, Chicago, Ill., 600 pp. (1973).

Moyer, D. D., Multiple-Purpose Land Data Systems, *Monitoring Foreign Ownership of U.S. Real Estate, Vol. 2*, U.S. Dept. of Agriculture, Washington, D.C. (1979).

National Aeronautics and Space Administration, *High Altitude Perspective*, NASA SP-427, Washington, D.C. (1978).

National Conference of Commissioners on Uniform State Laws, Uniform Simplification of Land Transfer Act, Chicago, Ill., 1-312, p. 37 (1977).

North American Institute for Modernization of Land Data Systems, *Proceedings of the North American Conference on Modernization of Land Data Systems (A Multi-Purpose Approach)*, 461 pp., Washington, D.C. (1975).

North Carolina Dept. of Administration, Keys to the Modernization of County Land Records, Land Records Management Program, Raleigh, N.C. (1981).

Office of Policy Development and Research, Report to Congress on the Need for Further Legislation in the Area of Real Estate Settlements, U.S. Department of Housing and Urban Development, Washington, D.C. (1981).

Pedowitz, J. M., Critique of the uniform simplification of land transfers act, *Real Property, Probate and Trust J. 13*, 696, 729 (1978).

Roberts, W. E., Software purchase, exchange, and transfer, in *Computers in Local Government: Finance and Administration*, Auerbach Publishers, Inc., Pennsauken, N.J. (1980).

Smith, D. E., Spaceborne Ranging System, in *Proceedings of the Ninth Geodesy/Solid Earth and Ocean Physics (GEOP) Research Conference*, Dept. of Geodetic Science Rep. No. 289, The Ohio State U., Columbus (1978).

Smith, R., Delaware Valley Regional Planning Commission, Philadelphia, Pa., private communication (1982).

Thompson, M. M., *Maps for America*, U.S. Geological Survey, Reston, Va. (1979).

U.S. Department of Agriculture, *Monitoring Foreign Ownership of U.S. Real Estate: A Report to Congress*, Washington, D.C. (1979).

U.S. Department of Housing and Urban Development, *American Land Title Recordation Practices: State of the Art and Prospects for Improvement*, Washington, D.C. (1980).

U.S. Department of Housing and Urban Development, *Profiles of the Land Title Demonstration Projects*, Washington, D.C. (1981a).

U.S. Department of Housing and Urban Development, *Summary of the Research Findings from the Land Title Demonstration Projects*, Washington, D.C. (1981b).

U.S. Department of Transportation, Federal Highway Administration, *Reference Guide Outline: Specifications for Aerial Surveys and Mapping by Photogrammetric Methods for Highways*, prepared by the Photogrammetry for Highways Committee, American Society of Photogrammetry (1968).

White, M. S., Technical reports and standards for Multipurpose Geographic Data Systems, presented to the Institute for Modernization of Land Data Systems (available from author at Bureau of the Census, U.S. Department of Commerce, Suitland, Md.) (1982).

Workshop on the Spaceborne Geodynamics Ranging System, IASOM TR 79-3, Institute for Ad-